DIGITAL ELECTRONICS

A Workbench Guide to Circuits, Experiments and Applications

ROBERT C. GENN, JR.

Illustrated by E. L. Genn

PRENTICE-HALL, INC.

ENGLEWOOD CLIFFS, NEW JERSEY

Prentice-Hall International, Inc., *London*
Prentice-Hall of Australia, Pty. Ltd., *Sydney*
Prentice-Hall of Canada, Ltd., *Toronto*
Prentice-Hall of India Private Ltd., *New Delhi*
Prentice-Hall of Japan, Inc., *Tokyo*
Prentice-Hall of Southeast Asia Pte. Ltd., *Singapore*
Whitehall Books, Ltd., *Wellington, New Zealand*

Library of Congress Cataloging in Publication Data

Genn, Robert C.
   Digital electronics.

   Includes index.
   1. Digital electronics.   I. Title.
TK7868.D5G46   621.3815   81-16837
ISBN 0-13-214163-9   AACR2

Printed in the United States of America

How This Book Will Put You on the Road to Success with Digital Electronics...

This practical reference work not only provides you with a firm foundation in digital electronics, but will give you lots of "hands-on" experience working with actual digital IC's, assembling logic circuits— and even building a variety of very helpful digital test instruments.

It is, indeed, a "workbench manual" offering complete guidance in the form of clear, simple, step-by-step instructions for 24 digital experiments and projects. You'll soon find the emphasis is placed on doing, rather than abstract theory. Every chapter includes helpful tips that will greatly simplify project building, cut costs and, in many instances, even help you reduce power requirements. All sections are fully illustrated with a wide range of charts, checklists, and work diagrams. In addition, Appendix A contains pin-outs for all of the most used TTL IC's, and Appendix C is a list of critical TTL and CMOS procedures—especially important for all readers.

Because of the strong, constant emphasis on practicality, you will acquire the expertise you need to work successfully with practically any digital circuit—using straight off-the-shelf, inexpensive solid state components. The various chips covered in detail in this book are all presently available.

New uses are being discovered every day, and the chips we'll use are found in electronic games, digital clocks, digital test equipment, medical instruments, process monitors, communications equipment, surveillance and security systems, as well as a large variety of other applications, including appliances, automotive and industrial controls. The extraordinary range of practical guidance offered by this book will help make your workbench a far more productive place, where you can immediately gain the essential

5

knowledge that is so important in servicing all types of digital systems. You'll discover that all the solutions to problems the digital technician experimenter is most likely to encounter are included. For example, how to get started on any project—how to select the most efficient digital components—how to interface your projects with other electronic or mechanical systems and how to identify and quickly correct defects in digital circuitry.

As all of us know, there is just no substitute for experience on any job. You can get this invaluable experience by working with digital circuits, right in your own shop, simply by completing the interesting and productive experiments described in this book. Every experiment will help you quickly gain that all important confidence and "know-how" needed to become a successful digital electronic technician.

Each chapter stresses the *practical* aspects, and describes inexpensive, simplified techniques and procedures that you will be able to use for years to come. To help you make fast repairs as easily as possible, all chapters show you exactly what to do in case you run into trouble. For example, Chapter 3 not only provides complete instructions on how to test and troubleshoot with digital test equipment, it also shows you how to build three of the most important pieces of test gear: a logic probe, pulser and clip. As another example, Chapter 9 contains descriptions and pin connections for various LED displays and shows you how to drive them—common cathode or common anode. Chapter 10 answers questions such as "Should I use a multiplexed or non-multiplexed system to drive my LED displays?"

Never before has the electronic technician faced as many real opportunities as he does today. Better pay, greater security, and more respect from your peers—they all go to those who take the time to keep up with their field, and every chapter that follows will provide you with the digital techniques that are so essential to your technical survival. Best of all, this book enables you to acquire a state-of-the-art expertise, with practical workbench experience right at home in your spare time.

Robert C. Genn, Jr.

CONTENTS

8. How to Select and Use Memory IC's *(cont.)*

9. Using Decoders/Drivers, Priority Encoders and LED
Displays in Digital Systems **223**

10. Practical Applications Using Data Selector
Multiplexers and Demultiplexers **252**

Other Books by the Author:
Practical Handbook of Solid State Troubleshooting
Manual of Electronic Servicing Tests and Measurements
Practical Handbook of Low-Cost Electronic Test Equipment
Workbench Guide to Electronic Troubleshooting

CHAPTER 1

A Practical Guide to Today's Digital IC's

Working with digital circuits does not have to be difficult and time-consuming; not if you have a knowledge of the individual **IC'S** and the logic family associated with them. This chapter offers an introduction to modern integrated digital circuits and includes pinouts, characteristics, and applications for all common **IC's** that are, in most cases, readily available on the market.

Each section includes valuable in-shop aids needed by every electronics technician/experimenter. You'll find simple ways to analyze many complex circuits and how to check inputs and outputs for various types of digital **IC's**. The skills you acquire from this chapter will save you time in building and troubleshooting digital circuits, since you will know what to expect from each logic circuit.

Basic Operating Principles of TTL IC'S

Today's electronics technician has available to him a previously undreamed-of assortment of hardware for his digital projects. First, there were vacuum tubes, then transistors, and now, integrated circuits **(IC's)**. The IC ushered in the present age of microelectronics by offering a single, thin semiconductor wafer (often called a *chip*), on an entire circuit, including diodes, transistors, resistors, capacitors, and internal "wiring." Of course, it's impossible even to think of replacing any of the internal electronic components and, therefore, we all must think of an **IC** as a so-called "Black Box." For our purpose, we can think of this black box as a device that will perform a certain electronic function *if* we are able to make it operate correctly. In order to do this, you must realize that digital circuits are electronic circuits that require numerical and logical data (electrical pulses) for their operations. Let's use the widely available 7400 series integrated circuits as an example of how digital **IC's** operate.

The 7400's use transistor-transistor logic (**TTL** or **T²L**), which refers to the type of electronic components used by the manufacturer during construction of the **IC**. As an example, Figure 1-1 shows a basic schematic of the internal components of a single **TTL NAND** gate. **AND** and **NAND** gates are thoroughly explained in Chapter Two.

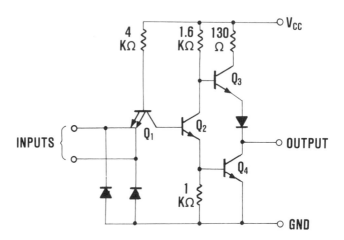

Figure 1-1: Schematic diagram showing the internal components of a 7400 TTL NAND gate IC (single gate)

Referring to Figure 1-1 and starting at the inputs, the diodes connected from the emitters of Q_1 are input protection against high-voltage transients. Transistor Q_1 has multiple emitter inputs and performs as an AND gate that is directly coupled to Q_2. Transistor Q_2 serves as an inverter, producing a **NAND** gate. The diode between Q_3 and Q_4 prohibits the two transistors from conducting at the same time. If Q_3 conducts, the output is a logic high and, if Q_4 conducts, the output is a logic low (i.e., one voltage is respectively higher or lower than the other). What we end up with is a **NAND** gate with a so-called "totem-pole" output. Another important thing is that the device has certain *fan-out* capabilities, or a single output point that can feed many gates of the same IC family. For example, in most cases, the output can drive ten other gates or devices.

Inside the 7400 **IC**, there are four of these separate, identical circuits (the schematic is shown in Figure 1-1). In this case, the **IC** is called a *Quadruple 2-input NAND gate.* The schematic shown in Figure 1-1 may be represented by a logic symbol like the one shown in Figure 1-2 (A). Now, when the manufacturer combines the gates into a single package and uses logic symbols, the information will be presented as a drawing showing **IC** pin numbers. See Figure 1-2 (B).

(A)

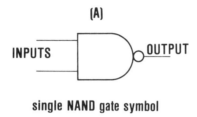

single NAND gate symbol

(B)

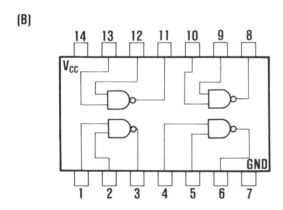

Figure 1-2: Quadruple 2-input NAND gate. (A) shows a single NAND gate symbol. (B) shows an entire package with the 7400 IC pin numbers. Normalized fan-out for each output is 10

The gate (any gate) is an electronic circuit having two or more inputs and one output. The output depends on which combination of logic signals you inject into the inputs. For example, the **IC** shown in

Figure 1-2 requires a logic high input at both input terminals (for instance, pins 1 and 2) to insure a logic low level at its output (pin 3).

CMOS, MOS, PMOS and Other IC Family Groups

To troubleshoot and build experimental digital circuits effectively, a firm understanding of various **IC** input and output characteristics for the logic family you're working with is required. Figure 1-3 shows the input and output circuitry associated with **PMOS, CMOS** and **MOS** logic families. See Figure 1-1 for information pertaining to the **TTL** family. In case you are not familiar with the term MOS, it stands for *metal oxide semiconductor.* The electronics switching device used to make **MOS, CMOS,** and **PMOS** is the metal oxide type field effect transistor or, **MOSFET.**

When a circuit structure has both P-channel and N-channel **FET** devices, it is known as a complementary circuit and leads to the name *complementary metal oxide semiconductor* or, **CMOS.** The other logic family, PMOS, is using P-channel enhancement mode **FET's.**

The most common **IC's** used by technicians/experimenters are the **MOS, CMOS,** and **TTL** we have discussed up to this point. However, there are a couple more types of logic that will help your understanding of how circuits are used in digital **IC's.** These are:

Diode-Transistor Logic (**DTL**): Figure 1-4 shows a **DTL** **NAND** gate. The diodes produce the **AND** function. The transistor is the inverter needed to produce the **NAND** gate, which is the basic gate for **DTL.**

Resistor-Transistor Logic (**RTL**): The basic **RTL** gate is very simple; two transistors and three resistors. See Figure 1-5. On the other hand, you'll find **RTL** logic used for numerous applications such as flip-flops, encoders, decoders, counters, and so on. Just remember, **RTL** logic is a family of logic composed of simple resistance coupled transistor circuits.

Understanding Logic Gates And Buffers

Although we used a Signetics Corp. 7400 **TTL NAND** gate as an example in Figure 1-1, it should be pointed out that there are several **IC** manufacturers that produce the same gate and, in general, all use the same identification numbers. However, there may be more

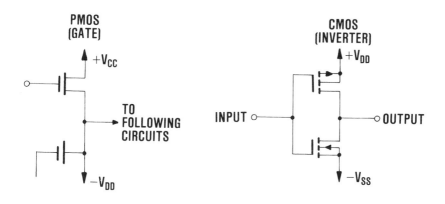

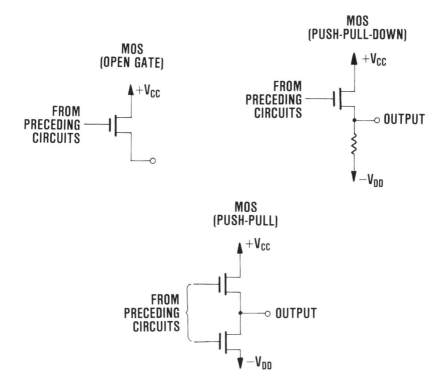

Figure 1-3:   Input and output circuitry for PMOS, CMOS, and MOS logic families. See Figure 1-1 for TTL input/output circuitry. You will find that *some* CMOS IC's have exactly the same pin configurations as the TTL 7400 IC's.

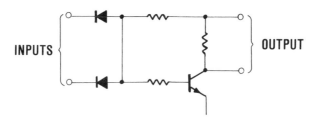

Figure 1-4:   DTL NAND gate

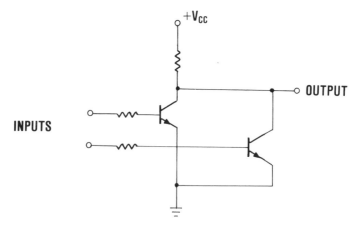

Figure 1-5:   Resistor-transistor logic (a basic NOR gate)

than one identification number for the same device. For example, you will find that an N7400 **IC** is, pin-for-pin, the same as an S5400. Taking it step-by-step, let's examine their identification system.

First, the letter **N** and the first two digits, 7 and 4. The letter **N** means that the **IC** is an *industrial* type, and the 7 and 4 mean the **IC** can be safely operated in a free air temperature range between 0⁰ and + 70⁰C. Now, what if you look up an S5400 **IC**? In this case, you'll find the **IC** to be exactly the same as the N7400, except that the **S** says the device is suitable for *military* use in a temperature range between -55⁰ and + 125⁰C.

Next, **TTL** logic can be subdivided into five families—regular, low-power, low-power Schottky, Schottky, and high speed. A letter between two sets of numbers (74 and 00, for example) may designate whether the **IC** is (H) high speed switching characteristics, (L) low-speed switching characteristics or (S) Schottky diodes. In the previous examples, there was no letter included (7400). This means the **IC** is standard. However, there is no relationship between the type of **IC** function and the last two digits. For instance, the 7400 is a quad **NAND** gate, while a 7490 is a decode counter.

Table 1-1 shows threshold-level comparisons for each of the five **TTL** families we have been discussing. In this table, both the absolute (maximum, minimum) and typical levels are shown. If you use this table of values during testing, you should use the absolute levels as your guide.

	LOGIC LEVEL	7400	74L00	74LS00	74S00	74H00
VOLTAGE LOW	TYPICAL	0.2	0.2	0.35	0.35	0.2
	MAXIMUM	0.4	0.4	0.5	0.5	0.4
VOLTAGE HIGH	TYPICAL	3.4	3.2	3.4	3.4	3.4
	MINIMUM	2.4	2.4	2.7	2.7	2.4

Table 1-1: Threshold levels for the five TTL logic families discussed in the text

From Table 1-1, you can see that for the standard **TTL** (7400), the low-level threshold is 0.4 volts, and the high-level is 2.4 volts. You might now ask, "What happens if I use a 74L00?" The answer is that the **L** identifies the **IC** as low power but, if you refer to the manufacturer's literature on this **IC**, you'll see that you must sacrifice speed to get this characteristic.

Going on over to the 74S00 **IC** shown in Table 1-1, you will notice that there is an **S** in the **IC** identifying number. As we have said, this means it is a high-speed switching device. The reason for the high speed is that the 74S00 has Schottky diodes, which have very fast switching characteristics. Incidentally, the 74S00 series is the fastest switching of the **TTL** devices.

The **IC** you choose to work with will depend on the job you want it to do. For instance, the speed and power requirements of your particular project will dictate which **TTL** device you need. As an example, you might use a 74LS00 if you want very little power drain and speeds as high as that of a 7400.

For a buffer, you would normally want a fairly hefty output. Therefore, and **RTL** buffer/inverter such as the one shown in Figure 1-6 possibly couuld do the job (depending on your requirements). Why this **RTL** circuit? Well, if you'll refer to Figure 1-5 you'll see that the **IC** has only two tansistors. On the other hand, Figure 1-6 shows two transistors in the output stage. Of course, this arrangement (push-pull) produces a greater output current capability—just what you would need for a buffer stage.

If you want to interface with an **MOS** device, lamp, or relay, you might choose a 7406 or 7416, which are, pin-for-pin, the same **IC**. This **IC** is a hex inverter buffer/driver with open collector high voltage outputs. Figure 1-7 shows a schematic of one of the inverters (each one in the **IC** package is exactly the same).

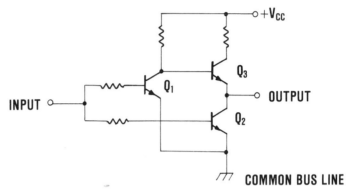

Figure 1-6:   Buffer/inverter with greater output current capability than the circuit shown in Figure 1-5

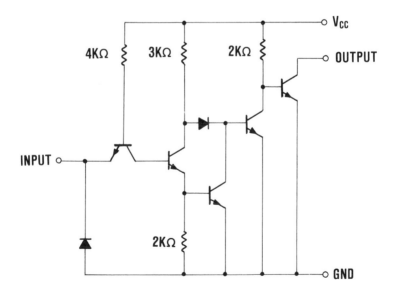

Figure 1-7:   One inverter schematic out of a hex (six circuits) inverter buffer/driver IC

One question that pops up here is, "If the 7406 and 7416 are the same, pin-for-pin, why couldn't you simply interchange one for the other?" Maybe you can, and maybe you can't. The secret is the recommended output voltage. The 7406 recommended output is 30 volts and the 7416 recommended output is 15 volts. Also, there are current capability differences between S and N types. What this boils down to is that you must have a description of all **IC** functions before interchanging one for the other, even though they are the same, so far as pin numbers are concerned. One other reason is that it's very possible there will be a slight difference in the **IC** package if you compare an **S** and **N** type.

Arithmetic Elements

Another **IC** in the **TTL** series is the 7480. This device is a gated full adder. Basically, an adder is an arrangement of logic gates that adds two binary digits (described in Chapter 2) and produces sum (Σ) outputs and inverted carry outputs ($\overline{\Sigma}$). The most basic step in performing arithmetic by **IC's** is addition. The 7480 **IC** uses diode-transistor logic (**DTL**, see Figure 1-4) for the gated inputs, and transitor-transistor logic (see Figure 1-1) for the sum and carry outputs. See Figure 1-8 for a schematic diagram of the 7480 gated full adder.

The binary adder is the most versatile digital circuit for performing arithmetic operations. As has been explained, the adder is capable of addition. However, it is also capable of subtraction, multiplication and division. Before we try to understand how a binary adder works, it's best to first acquire a knowledge of the binary numbering system and logic diagrams. These subjects are covered in Chapter 2.

Character Generators

A character is a letter, number, or symbol and is usually found on the face of the cathode ray tube (CRT). Next, what is a character generator and where would you most likely find one of these devices? Basically, the character generator is an **IC** that receives an input signal (some form of alphanumeric code) and converts binary data into electrical voltages that manipulate a CRT electron beam to trace out characters on the face of a CRT. In general, you'll find that all microcomputer video terminals are using some type of

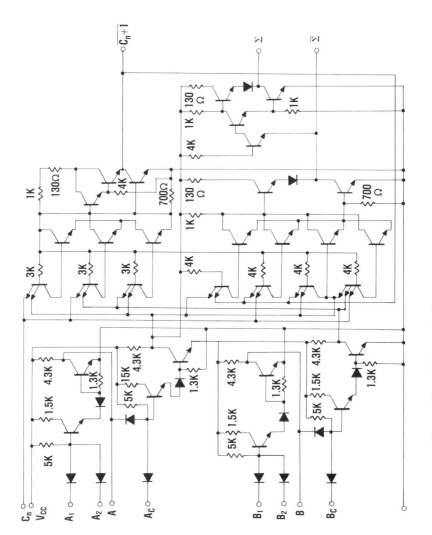

Figure 1-8:   Schematic diagram of a 7480 gated full adder IC

character generator. Typically, one of these CRT displays is capable of showing 960 characters at one time (12 rows with each line being made up of 80 characters, i.e., the letters, numbers, or symbols you see on the CRT).

Decoders/Display Drivers

One of today's most popular decoder/display drivers is the seven-segment decoder/driver. You'll find these IC's used with digital clocks, watches, calculators, digital voltmeters, and many other electronic devices from medical instruments to space craft. Figure 1-9 shows the pin configuration of a Signetics 8T04/5/6 seven-segment decoder/driver with an accompanying seven-segment standard layout. The IC pins labeled a, b, c, d are fed with digital input signals (binary coded decimal) and A, B, C, D, E, F, and G are seven-segment outputs, in accordance with the standard shown. Auxiliary inputs are ripple blanking input (RBI) and ripple blanking output (RBO). These are used to suppress leading and trailing edge zeros in multidigit displays. The LT input is a test pin and allows the device to be checked at any time.

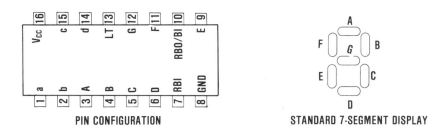

PIN CONFIGURATION   STANDARD 7-SEGMENT DISPLAY

DECIMAL DISPLAY

Figure 1-9: Seven-segment decoder/driver pin configuration and standard seven-segment layout, with associated decimal display

The various segments needed to create the ten decimal digits are as shown in Figure 1-9. However, if you are working with this type **IC,** it is important that you know that various decoders have different electrical characteristics. For example, the 8T04 has "active low" (logic activating level), high-current sink open collector outputs for driving indicators directly. The 8T05 has "active high" outputs with internal pull-up resistors to provide sufficient drive current to

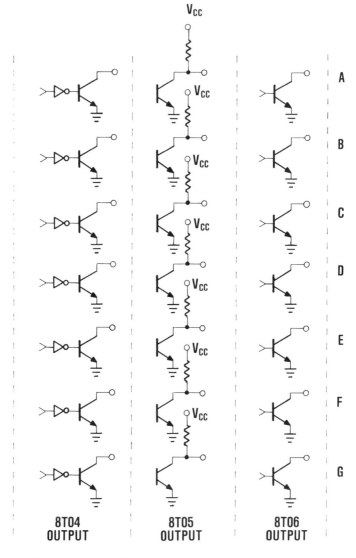

8T04 OUTPUT   8T05 OUTPUT   8T06 OUTPUT

Figure 1-10: Examples of different electrical characteristics of IC's (8T04/5/6) with the same pin configuration

discrete transistors, SCR's and other interface elements. The 8T06 also offers "active high" outputs, but these are of open collector type so that you have a variety of current source applications. Figure 1-10 shows the output diagrams of the 8T04/5/6. You'll find practical applications, using BCD-to-Decimal decoders and similar devices, in Chapter 9.

Counters

A counter **IC** is a circuit that counts input pulses. However, there are several different types of counters. Serial (also called *ripple-through counter*), ring, and asynchronous are typical examples of the various names you may encounter. Furthermore, there are a variety of **IC** counters in each type. For instance, an asynchronous **MSI** counter/storage element may be any of the following: standard 8280/81/88, 8290/91, 8292/93, or Schottky 82S90/91.

Asynchronous counters (an **IC** in which the speed of operation is not related to any frequency in the system) have a couple of advantages over other counter configurations. You can keep power consumption to a minimum, utilizing the fact that each stage (flip-flop circuit) does not have to be operated at the incoming frequency. A synchronous counter is one in which the performance of the sequence of operations is controlled by equally spaced trigger pulses (clock signals). On the other hand, in a system such as a binary ripple counter (in an asynchronously controlled counter, the clock signal is derived from a previous stage output. See Figure 1-12.) each stage (flip-flop) operates at one-half the frequency of the preceding stage. You'll find that this type is easier to work with (compared to synchronous counters) and power consumption is reduced.

To understand **IC's**, one must have some idea how a so-called *flip-flop* circuit (FF) works. Figure 1-11 shows a basic flip-flop circuit. This circuit can remain in one of two possible states for an indefinite time, thereby acting as a storage for a single pulse (bit) of information. The information held is either a yes or no, 1 or 0, etc.

The characteristics of this simple stage are: if input #1 is made high (for example, a higher voltage than a low state, often identified by the number 1), output #1 will be inverted, or a low state (0); making input #2 high (1), output #2 will be low (0) but, because the two stages are feeding into each other, you'll always find that one stage *output* (either #1 or #2) will be in the same state (either high or low) as the *input* of the opposite stage. To change from one set of conditions

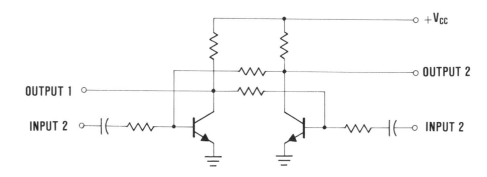

Figure 1-11: A basic flip-flop circuit (FF) that can be used as a counter/storage element

to the other only requires a change of one of the inputs, using a short electrical pulse (1 or 2).

Starting with the simple basic circuit just described, flip-flops have been used in the design of many counters. However, there are simpler ways that you can use to accomplish the same job (these will be discussed in later chapters). Combining flip-flops and gates can result in a counter/storage element such as the 4-bit binary ripple counter (8281) shown in Figure 1-12.

The **IC** shown in Figure 1-12 has both a strobe input (S_d) and a reset input (R_D). While the strobe line is activated (input signal at logic level 0, in this case), the clock input will have no effect on the flip-flop. Using the strobe input (injecting a reset pulse), it's possible to reset a flip-flop, or clear to zero asynchronously. When you are working with a counter such as this one, it is necessary to consider strobe and reset release times if it is desired to count after a reset or strobe operation—typically, on the 8280/1 **IC's,** 30 ms for strobe and 50 ms for reset.

Interface Elements

You have to read the data sheets very carefully if different types of logic (**MOS, TTL, DTL,** etc) are to be successfully used together (interfaced). You should check the operating voltages, maximum and minimum logic levels, DC input current, output/input characteristics and noise immunity. There are lots of different **MOS** technologies, and each takes one of the interface circuits shown in Figure 1-13.

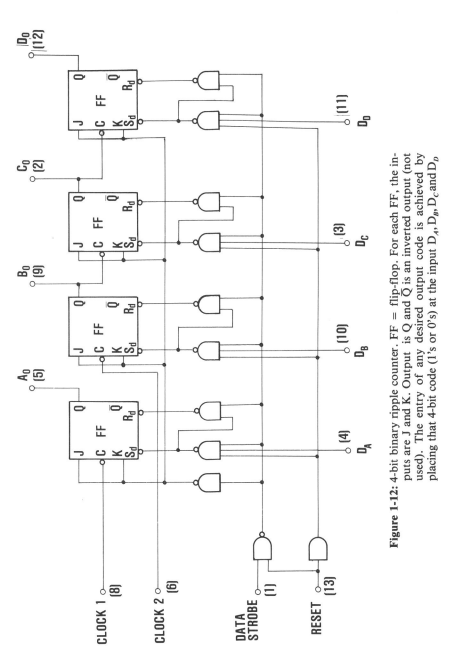

Figure 1-12: 4-bit binary ripple counter. FF = flip-flop. For each FF, the inputs are J and K. Output is Q and $\overline{Q}$ is an inverted output (not used). The entry of any desired output code is achieved by placing that 4-bit code (1's or 0's) at the input D_A, D_B, D_C and D_D

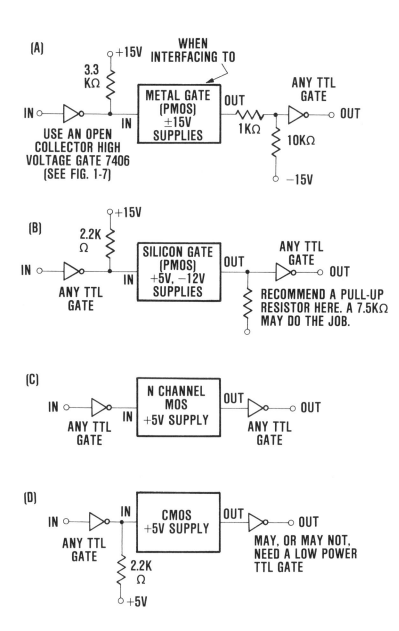

Figure 1-13: Interfacing MOS devices and TTL gates (A) using a 7406 to drive a metal gate (±15V). (B) silicon gate circuits (generally, a +5V and −12V supply), resistors shown will usually interface. (C) and (D) N-channel and CMOS will normally interface any two TTL gate IC's (± 5V supply is an indication of the type devices).

One thing that makes life at the workbench much simpler is that, in most cases, you can tell the type of **MOS** device by the supply voltage used or recommended. As an example, if the supply voltage is ± 15 volts, the chances are the **IC** is an older type **TTL** or **CMOS** and needs an open circuit **TTL IC** that can withstand 15 volts. Where can you get a suitable **TTL IC**? Easy. Referring to Figure 1-7, you'll find it to be a hex inverter buffer/driver with open collector high voltage outputs. This **IC** has **TTL** inputs and can do an excellent job of interfacing with **MOS**.

In general, you'll find that **TTL** devices used to drive **CMOS** use one of three possible output circuits, namely resistor pull-up, open circuit, or transistor (active pull-up). For an example, see Figure 1-3. However, it's very possible, as has been shown in the last few paragraphs, that you can get by with nothing but a pull-up

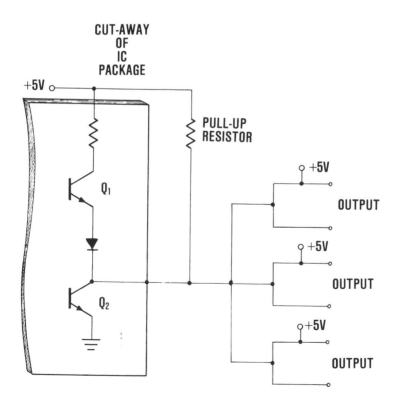

Figure 1-14: Using a pull-up resistor to interface a TTL IC to a CMOS IC

resistor (the open collector **IC** becomes a resistor pull-up when you add an external **IC**). The idea of adding a pull-up resistor is to increase the output voltage of the **TTL** device you are trying to interface with. See Figure 1-14.

Referring to Figure 1-14, you'll notice that the pull-up resistor is in parallel with the transistor Q_1 and the diode. What this does is make the output voltage across the pull-up resistor equal to the sum of the voltage drop of the transistor and diode, i.e., increases the output voltage. Incidentally, if you are wondering what the rest of the schematic in Figure 1-14 looks like, see Figure 1-1.

When combining logic circuits (**IC's** etc.) to build entire systems, it is often necessary to drive several stages. This brings up "fan-in" and "fan-out." Fan-in is usually defined as the number of inputs that can be connected to a certain **IC**. Fan-out is telling you the number of parallel loads that you can connect to a certain **IC** with satisfactory results, *provided* the loads are of the same logic family as the driver stage.

Another **IC** that is manufactured to be used as an interface is the dual 2-input **NAND** high voltage **TTL** interface gate 8T18. See Figure 1-15, which shows a circuit schematic of one-half the 8T18 **IC**. The unit contains two equal circuits, as shown symbolically in the package diagram.

The basic gate shown, works from two power supplies (V_{CC_1} and V_{CC_2}). The input circuits use a high voltage supply (V_{CC_2}) between 20 and 30 volts and the output transistors require a standard 5 volt power supply (V_{CC_1}). The output circuit features active pull-up and pull-down, providing a low impedance driving source in both high and low output states. This device is particularly suited for driving the high capacitance loads encountered in fan-out and line-driving applications. Practical applications and fan-in/fan-out problems will be thoroughly analyzed in the following chapters.

Memories

In the most basic sense, a memory circuit is any digital circuit that can store a voltage pulse (either logic high or logic low). For example, the flip-flop circuit shown in Figure 1-11 is a basic memory device. A group of similar flip-flops can be combined to store a group of bits of information (1's and 0's). However, what we are interested in at this point are memories in IC form. In particular, read

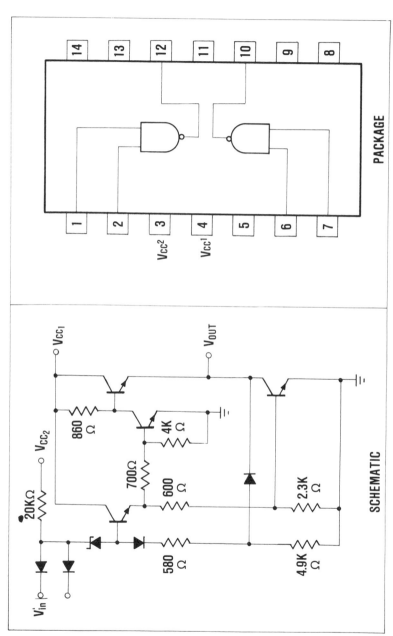

Figure 1-15: Dual 2-input NAND high voltage to TTL interface gate. This IC uses different pin configurations for different packages

only memories (ROM), random access memories (RAM), programmable read only memories (PROM), and erasable PROM's.

The 7488 **IC** is a 256-bit read only memory and the 7489 is a 64-bit read/write memory. Let's just examine the characteristics of the 7488 **ROM** and see how it's possible to make it work for you. First, the 7488, like all other **ROM's**, is a memory from which data can be read out repeatedly but, once it's programmed, it cannot have any other information entered, i.e., it cannot be changed or reprogrammed. The basic **ROM** may be programmed by the manufacturer. In fact, the customer usually will specify patterns (information to be stored) for the **ROM** by completing an instructional order blank. Figure 1-16 includes pin configurations and a logic diagram for a 7488 **ROM.**

The 7488 **ROM** has five address select inputs, A0-A4, and eight outputs, B0-B7. The purpose of the memory enable input (C_s) is to override the address select inputs and, when it is at a logic high state, all outputs are inhibited and will be at logic 1 state. The device is full **TTL** or **DTL** compatible. Obviously, **ROM's** have drawbacks. However, there are pin-for-pin replacements for the 7488. One of these is the 256-bit bipolar field programmable **ROM** 8223, which may be programmed to any desired pattern by the user. This may be done by using a fusing procedure, which is normally furnished by the **IC** manufacturer in their sales catalog and other technical information.

Although the **ROM** is useful for such things as electronic reference tables (such as trigonometric tables, etc.), and for character generators in video display systems, these applications do have obvious disadvantages for the home experimenter. In general, you'll find that **ROM** and **RAM** are basically the same technology in that both are memory devices. However, the **ROM** usually has a built-in coded message whereas, as the name implies, the **RAM** is a memory in which information may be stored and withdrawn in a time interval that is independent of the memory cell address. See Figure 1-17, which shows the pin configuration and logic diagram for a 64-bit read/write memory 7489.

The 7489 is a **TTL** 64-bit **RAM** organized as 16-words of 4-bits each. Words are selected through a 4-input binary decoder when the chip select input (C_E) is at logic 0. Data is written into memory when read enable (R_E) is at logic 0, and read from the memory when R_E is at logic 1.

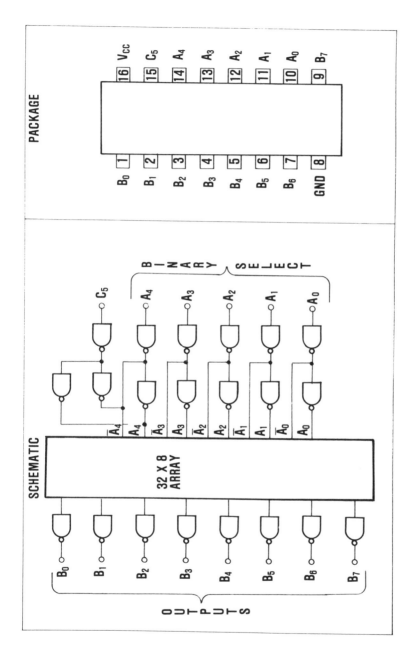

Figure 1-16: 256-bit read only memory (7488) logic diagram and IC pin configuration

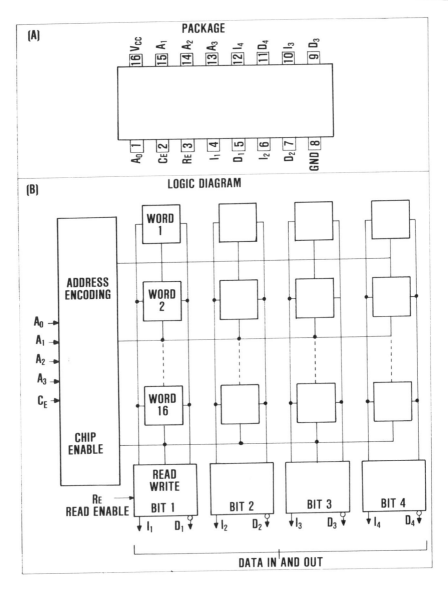

Figure 1-17: 64-bit read/write memory 7489. (A) is package showing pin configuration and (B) is the logic diagram

Perhaps the most important of all the memory **IC's** are the **PROM** and **EPROM** (erasable **PROM**). The **PROM**'s are probably today's work horses in the digital electronics field and you should become familiar with these very useful devices. The programmable read only memory is a member of the **ROM** family. In fact, the only difference, as has been pointed out, is that you can program the **PROM** yourself. Hence, the **PROM** is also called a *field programmable* **ROM**.

One **PROM** that is **TTL** compatible is a 16-pin dual-in-line package 8223, as has been mentioned. At this point, let's see how the actual programming procedure would be done. First, certain internal junctions of the **IC** must be either closed or fused open, depending upon how the **PROM** is programmed. But, before this is done, one must determine what binary words should be programmed into which address location in the **IC**. Then select each address and program the bits (1's and 0's) of each word, one at a time, until the programming of the **PROM IC** is complete. Figure 1-18 shows a workbench setup that could be used to program a **PROM IC**. The actual step-by-step programming procedure is included in Chapter 8.

Just as there are closed or fused open junctions in a **PROM**, the **EPROM** uses static charges on **MOS** field effect transistors to

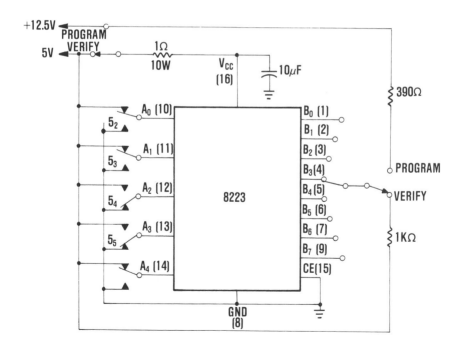

Figure 1-18: Manual programmer diagram that could be used to program a PROM 8223

achieve the same effects. The best part of all, whatever is placed into the memory can last for years, or be erased in a few minutes by special ultraviolet lamps. **PROM**'s and **EPROM**'s are available in many configurations. All you have to do is check the manufacturers' catalogs, or look in the back of almost any electronics magazine for the **PROM** you need for your application. Incidentally, in some cases the distributor will program the **IC** for you at no additional cost. *Note: A word of warning. Always check cost of item and mailing costs before ordering any electronic component!*

Multiplexers

A multiplexer is a switching device that can select one of several inputs and connect it to a single common output. For instance, if data has to be selected from two registers, it can be done simply with the 8233 multiplexer. An example wiring diagram is shown in Figure 1-19.

To understand the function of multiplexers such as the 8233, it's best to view them as a 4-pole, 2-position switch. By placing the correct logic level into S_0 and S_1 (select inputs), either register 1 or 2,

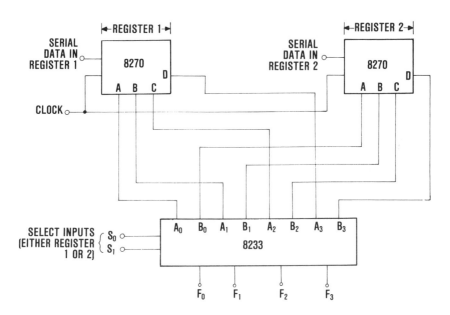

Figure 1-19: An example of using a 2-input 4-bit digital multiplexer (8233) to select data from two registers

or no outputs, may be made to appear at outputs F_0 and F_1, F_2, and F_3. Other fairly common multiplexers are the 74153, which can serve as a double-pole 4-throw switch, the 74151, and 8:1 multiplexer, and the 74150, a 16:1 multiplexer. This 24-pin **IC** has four data select inputs that will cause it to select one of 16 points.

Registers

Basically, a register is several flip-flop circuits used for storage information. However, not only must the device store data but it usually has to move the data from one location to another on command. Two inexpensive registers are the 7495 (4-bit shift register) and the 7486 (5-bit shift register). Figure 1-20 shows the pin configurations for both.

Shift registers are made up of individual stages (flip-flops). Each stage can store one *bit* of information called a *binary* 1 or 0, and usually corresponding to a *yes* or *no* command. Four bits together can represent a decimal number, etc. The contained information may be moved (or shifted) from stage to stage. This shifting is called *clocking* and one or more clocks are involved in completing the shifting operation.

The 7495 **IC** shown in Figure 1-20 is a 4-bit right-shift/left-shift **IC**. The 4-bit designation means the register contains 4 flip-flops. More important, the number of flip-flops determines the amount of data per unit. It's not only important how many stages a cetain **IC** has, but how you can get at the stages. For instance, if the device is a serial-in/serial-out register, it gives you input only to the first stage and final output of the last stage. Sometimes you'll hear this **IC** called a serial register. An 8-bit 74164 is a typical example of a serial register. If you are working with this **IC**, there is no access to the intermediate stages.

Data can be moved from one register to the other by serial shifting. But, a faster method of transfer is called *parallel.* A serial-in/parallel-out gives you the outputs of all stages, including the last one. The 8-bit 74164 is a typical **TTL** example. Another approach is the parallel-in/serial-out register **TTL** 7494. This **IC** let's you simultaneously load all the stage but then steps in contents out as a serial-bit string.

One of the things you'll find in manufacturers' catalogs when you look up register **IC**'s is that some of them are listed as being *dynamic* (for example, Signetics 2505), and some are listed as *static shift registers* (for example, Signetics 2533). Figure 1-21 shows two

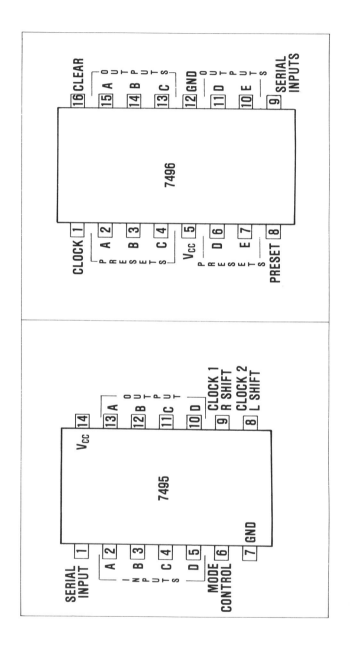

Figure 1-20: 7495/6 shift register IC pin configurations

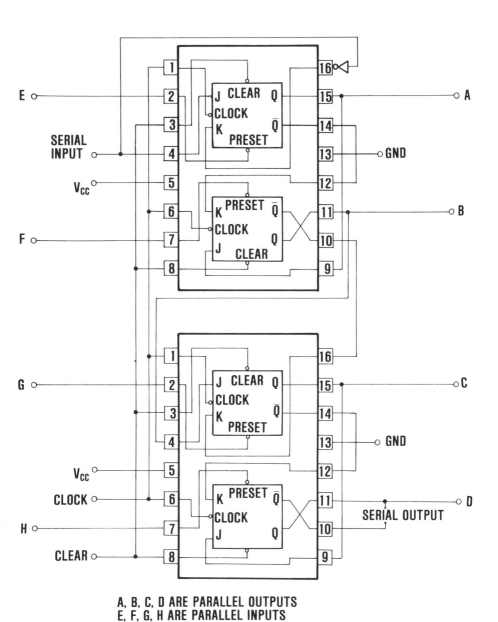

A, B, C, D ARE PARALLEL OUTPUTS
E, F, G, H ARE PARALLEL INPUTS

Figure 1-21:   Shift register using J-K flip-flops (7476 IC's). Another low-cost
IC that could be used in this circuit is the 74107.

7476 dual J-K flip-flops that can be connected to form a shift register. They will keep data so long as you apply power, and are called *static registers.*

The transformation of information in any shift register has to be done by two stages at a time. In such a case, as shown in Figure 1-21, we have a temporary (often called *master*) and a final (often called *slave*) storage within each J-K flip-flop's logic block. The sequence of operation is as follows:

1.   Isolate slave from master.
2.   Enter information from J and K inputs to master.
3.   Disable J and K inputs.
4.   Transfer information from master to slave.

Sometimes you don't need a full flip-flop. In such cases, you can use a capacitor for temporary storage. Figure 1-22 gives you the basic idea of how this is done. The setup is called a *dynamic shift register.* As always, one never gets something for nothing. The cost, here, is information holding time. Most true dynamic **MOS** shift registers will hold information for about one-tenth of a second, at best. In other words, if you don't get new information entered in that time, you'll lose all information. By the way, it makes sense if you think about it. All capacitors leak current, i.e., stored voltage will go to zero in a faily short time.

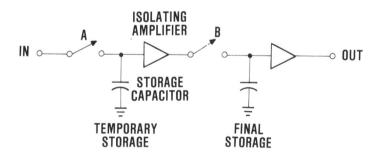

Figure 1-22:   Basic idea of a dynamic shift register

Figure 1-23 shows a better idea (generally all static **MOS** registers use this system). Using a capacitor for temporary storage and a flip-flop for final storage will produce a register with static performance for less cost. There are rules to remember when using one of these **IC**'s. One is that you must keep the clock line at a certain level during the static partt of the operation (during the holding time) and you must not get over the *maximum* allowable clock pulse width (time of duration) when you cause the stages to go through the dynamic transfer process.

In summary, we can say that, as in all **IC**'s, you have to ask a few questions before using the device. For example:

1. Is the register *static* or *dynamic?*
2. How do you interface a register **IC**, or any other **IC**, with **TTL** or other logic?
3. What kind of clock signals (if any) are needed and how many do you need?
4. If it's a register **IC**, can the circuit recirculate by itself?
5. Does the register **IC** have write or read enables so you can use it with add-on register **IC**'s?

So you can instantly answer your daily on-the-job questions, we will take a look at all these important questions in much more detail in Chapter 6.

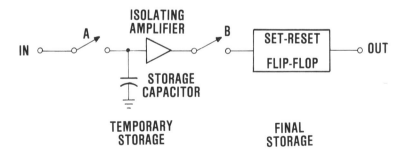

Figure 1-23: Simplified block diagram of a typical MOS shift register that has static shift register capabilities under certain conditions (see text)

CHAPTER 2

Using Digital Fundamentals in Digital Applications

Whether you design, repair, or maintain digital **IC** equipment, the digital reference data in this chapter will help you. On just about every page you'll find one or more illustrations; for example—troubleshooting aids, function tables, logic diagrams, **IC** structures, and pin configurations, to help clarify the explanations, systems and devices.

Included are actual logic diagrams of the four basic logic functions: **AND, OR, NAND,** and **NOR,** plus the operating characteristics of the three basic flip-flops, **RS, D,** and **J-K.** All these functions are important to you because they are the backbone of modern digital electronics and an understanding of them will help boost your earning power. Each page in this chapter is geared toward saving you time and effort in every phase of your work. You'll find pin numbers and working specifications for many **IC**'s. You'll also learn how to effectively use the tools of Boolean algebra to quickly identify a trouble; how to work with logic equations and truth tables, and techniques essential for troubleshooting or designing your own logic circuits. All examples in the following pages focus on the truly practical situations that you as a technician/experimenter could encounter on your workbench.

40

Sample Uses of Numbering Systems

Digital circuits use only two "states," on and off, or 1 and 0, usually called *binary digits*. These binary digits, referred to as *bits* are represented by a voltage on the output or input of a digital device. If the voltage changes from a high to a low, it usually means that the output or input changes from +5 V (high) to 0 V (low). The opposite is true for a change from low to high; i.e., the state changes from logic low (0) to a logic high (1). In many actual circuits, low may not go to zero but only drop from about +5 V to, maybe ±1.2 V. The *change of state* is the actual signal pulse and, if it's large enough, it will cause the following digital circuit to change its state.

Each electrical pulse (binary digit) has a negative-going edge and a positve-going edge. The digital circuit you're working with must detect the transition, and it is the edge of the transition that is used by the circuit to change its logic level (go from 1 to 0, etc.). Figure 2-1 illustrates the most interesting nomenclature and definitions used in connection with digital bits of information, and can be used during troubleshooting.

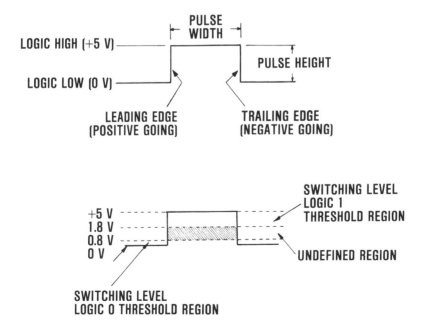

Figure 2-1:   Terms used when referring to digital circuit pulses (bits). This is called *positive logic,* the 1 bit being 5 volts and the 0 bit being near 0 volts.

The lower part of Figure 2-1 (B) shows a pulse for the operation of a digital circuit. The shaded area shown (0.8 V to 1.8 V) is sometimes called the *undefined area* of the pulse. The sections of pulse above and below the threshold are considered to be the normal operation region; i.e., logic 1 (between + 1.8 V and + 5 V) and logic 0 (between + 0.8 V and 0 V). *Important to the troubleshooter:* The digital circuit the pulse is driving *may* or *may not* respond to the input, if the pulse amplitude falls anywhere in the shaded area. Threshold levels for **TTL** logic families are shown in Table 1-1, in Chapter 1.

For most **MOS/CMOS IC**'s that are **TTL** compatible, the low-level putput is 0.05 volts (this is maximum logic 0, assuming no noise on the input with only capacitance loading), and minimum high level logic (1), under the same condition, is given as 4.5 volts. *Notice,* if we represent the 1 bit with a —4.5 volt and the 0 bit with a —0.5 volt, it is called *negative logic.* Remember, all variations in power supply potential, V_{cc}, will directly affect the threshold point.

1. Binary Numbering System

It will be helpful to you, when working with binary, if you can change from a binary number to a decimal number and vice versa. The simplest way I know is to use the chart shown in Table 2-1.

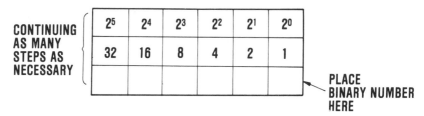

	2^5	2^4	2^3	2^2	2^1	2^0
CONTINUING AS MANY STEPS AS NECESSARY	32	16	8	4	2	1

PLACE BINARY NUMBER HERE

Table 2-1: Conversion chart for changing from binary to decimal

To use the chart shown in Table 2-1, start writing the binary number from the right-hand side of the bottom line. For example, let's say that you want to know what the binary number 101 (read from right to left) is in decimal. First, write the binary 101 directly under the conversion table, as shown. Next, simply add each and every number in column 2 that is directly above a 1 of the binary number you have written in line 3. In this case, you add the 1 and 4 (not the 2 because there is a 0 shown) and this completes the conversion. Binary 101 is equal to the decimal number 5.

How do you convert from decimal to binary? Simple. Using the number 5 again, place a 1 under 4 (2^2) and a 1 under 1 (2^0), and a 0 under all other numbers. Then read from the right-hand side of the bottom column. In this case, you're back to 101, the binary number for decimal number 5 (see Table 2-2).

	2^2	2^1	2^0	LINE 1
CONTINUING AS SHOWN IN TABLE 2-1	4	2	1	LINE 2 BINARY NUMBER
	1	0	1	LINE 3 WRITTEN IN

Table 2-2: Using Table 2-1 to convert the binary number 101 to the decimal number 5

What if you wanted to know what the binary number 100110 is in decimal? If you place all digits in the bottom column in Table 2-1, then add all numbers in the second column above each 1, you find that the decimal equivalent is 38.

The binary equivalent operations of addition, subtraction. multipilcation, and division are essentially the same as for decimal arithmetic operations. That is, once you know the rules of the game, it isn't very difficult to perform the operations. Because there are numerous mathematical text books on this subject, and because this is mainly a practical book, we will not purse binary arithmetic any further. However, there are other numbering systems used in digital work that you should at least have a nodding acquaintance with. These are the *octal* numbering system, the *hexadecimal* numbering system and, very important, *binary* coded decimal.

2. Octal Numbering System

The octal numbering system is based on the powers of eight, just as the binary numbering system is based on powers of two, and the decimal numbering system is based on ten. Referring to Table 2-1, you'll see that 2^3 is 8, and this is an important point because it brings out the fact that the octal system is a multiple of the binary system and conversion between the two is relatively easy. This is especially important to computer people because, when working with computers, it makes programming much easier.

Let's see how one would convert a binary number to a number in the octal system. Binary number 011 101 100 = ? octal. First, group the binary number into groups of three, going from right to left, and convert each group of three to its octal equivalent (use Table 2-1). You should come up with:

011	101	100
3	5	4 = 354 octal

3. Hexadecimal Numbering System

Another numbering system you may need is the hexadecimal (base 16). It, too, is a multiple of the binary system. To convert from binary to hexadecimal, you simply group bits into groups of four and then convert each group. However, we do run into a problem here because there is no single arithmetic symbol above 9. To get around the problem, the alphabetic symbols A through F are used. This means that in the hexadecimal system you'll find 0, 1, 2, 3, 4, 5, 6, 7, 8, 9, A, B, C, D, E, F, being used. As an example of what you might expect to find:

0110	1010	0101	binary
6	A	5	= $6A5$ hexadecimal

4. Binary Coded Decimal Numbering System

The next numbering system that you should be familiar with is binary coded decimal **(BCD)**. **BCD** code uses 4-bit groups to represent each of the ten decimal digits. It should be pointed out that a **BCD** code is not a basic numbering system such as are decimal, binary, and octal. It is simply a widely used system for coding a decimal. For instance, we would write the number 5 decimal as 0101 in **BCD** because, in **BCD**, numbers are always in groups of four. Now that we have a basic understanding of numbering systems, etc., let's take a look at binary words.

Binary Words

As you know, a bit is a digit in binary language. Binary numbers are also grouped into binary words. For instance, a 5-bit binary number may be referred to as a *5-bit word, computer word, data word,* or just, *word.* Figure 2-2 shows the pin configuration for

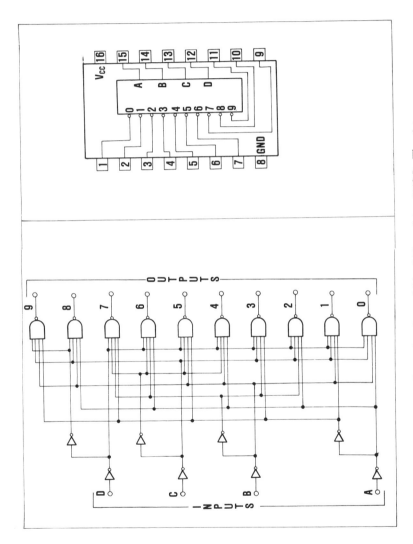

Figure 2-2: Pin configuration for a 5-bit shift register (7496). The data-per-unit is often referred to as a *word*; in this case, a 5-bit word

a 5-bit shift register **IC** (7496). Notice the outputs are pins 10, 11, 13, 14, and 15. From previous information in Chapter 1, we know this **IC** is made up using five flip-flops because, as you'll remember, the number of flip-flops determines the amount of bits that can be stored.

All flip-flops are simultaneously set to logical 0 state by applying a logical 0 voltage to the clear input. The common preset input is provided to allow setting of all flip-flops simultaneously, or one at a time. Also, since both inputs and outputs to all flip-flops are accessible, parallel-in/parallel-out, or serial in/serial-out operation may be performed. It should be noted that a 5-bit binary word can represent 32 different combinations or a 2-bit binary word can represent 4 different combinations. For example, Table 2-3 shows the maximum combinations of a 2-bit word.

DECIMAL	3	2	1	0
BINARY	11	10	01	00

Table 2-3: Maximum number of *combinations* that can be represented with a 2-bit word (4)

The highest *decimal number* that you can express, using a certain word, is the total number of combinations minus one. It's easier if you use the equation: decimal number $= 2^n - 1$. For example, the highest decimal number that can be expressed by a 4-bit word is:

$$2^n - 1 = 2^4 - 1 = 16 - 1 = 15$$

You can see this by referring to Table 2-1. Notice, to express the decimal number 16 as binary, you must use five binary digits (10000).

A combination of eight binary digits is commonly used to form a binary word in equipment such as microprocessors. Therefore, an 8-bit system is one in which eight binary digits are used to form a word. When all eight bits are used at once, it's often referred to as a *byte*. However, there are many time where less than eight bits are required to express a number. In an 8-bit system, four bits are generally called a *nibble*. It should be pointed out that you may find other amounts of bits referred to as a nibble. For example, there are quite a few computers that will accept either a 16-bit or 8-bit word. Anything less than a complete word may be called a nibble.

BCD to Decimal Decoder IC's

As has been pointed out, one of the most popular of the practical forms of decimal conversion is called *binary coded decimal* **(BCD).** Table 2-4 shows how each decimal digit can be represented by four binary digits.

DECIMAL DIGIT	BINARY CODE
0	0000
1	0001
2	0010
3	0011
4	0100
5	0101
6	0110
7	0111
8	1000
9	1001

Table 2-4: Binary coded decimal (BCD). In BCD, each digit of the decimal number is individually converted to a binary number, as shown

Some of the most inexpensive **IC**'s that are used with binary coded decimal signals are the 7441 and 7442, both **BCD** to decimal decoders. The difference between the two is that the 7441 is a decoder/driver and the 7442 is a **BCD** to decimal decoder, without driver circuits. The 7442 **IC** pin configuration and logic diagram are shown in Figure 2-3.

If you are working with an 8-bit system (such as a microcomputer), you'll find each byte can be viewed as containing two 4-bit

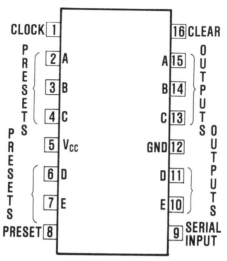

Figure 2-3: BCD to decimal decoder (7442). This IC decodes a 4-bit BCD number to one of ten outputs.

BCD numbers. However, equipment such as computers use internal codes (**BCD**) and external (user) codes. Two codes often used by the operator/programmer are hexadecimal and ASCII. Conversion between the various codes is done by decoders, encoders, and code converters. By the way, when you are working on computers and the like, it is important to remember that all computers do not use the American Standard Code of Information Interchange (ASCII). To put it another way, you may very well run into other types of keyboards or printouts. The good news is, inside all digital equipment, you'll find binary in use. This book will help you when working with all type **IC**'s, etc. But, if you are going to work on microprocessor systems, you'll also need operational instructions before the computer will perform any function. This information is provided by the manufacturer and is generally called an *Instruction Set.*

How Understanding Simple Boolean Algebra
Can Help You

Knowing how to effectively use the tools of Boolean algebra will help you with digital circuit analysis, design work, troubleshooting and repair, and selection of the appropriate **IC**. Because this book is written for real-world, hands-on experience, let's use practical examples. For instance, in a manufacturer's data book or applications notes you may find a Boolean equation that looks like this:

$$3 = (4 + 5) + (6 + 7)$$

where each number represents a pin on a particular **IC** gate. Mathematically, a Boolean equation usually is given in algebraic form, i.e., A, B, C, etc. However, at the workbench, all of us normally work with **IC** pin numbers so we'll use these rather than algebraic symbols.

Now, let's examine the Boolean equation and see how it can be used. First, the expression is telling you there are two **OR** gates being **OR** 'd by another single **OR** gate. Next, drawing the symbol for the first part of the equation (pin 4 + pin 5), we get Figure 2-4.

The second part of the equation can be accomplished using another **OR** gate exactly like the one shown in Figure 2-4. Combining the Two **OR** gates and adding a third (the + sign indicates we need a

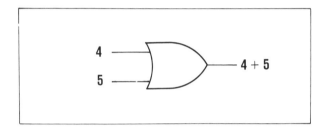

Figure 2-4: Symbolic drawing of an OR gate, showing the output to be the sum of the signal injected into pins 4 and 5

third), we end up with a complete *logic* diagram of an **IC** containing the necessary gates to complete the Boolean equation. This is shown in Figure 2-5.

Now that you have a general idea of what a Boolean equation can be used for, let's go one step farther and analyze an **IC** that uses two **OR** gates and one **AND** gate. Figure 2-6 shows an **IC** constructed using these three gates.

In this case (Figure 2-6), we need the Boolean equation for the two OR gates (3 + 4) and (5 + 6). Now, because the **AND** gate is doing the combining, the total equation becomes:

$$\text{output (pin 2)} = (3 + 4) \bullet (5 + 6)$$

(the symbol for multiplication ($\bullet$) is read as **AND** in Boolean algebra). The multiplication is the result of including the **AND** gate.

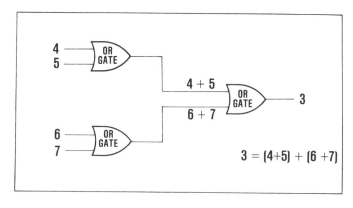

Figure 2-5: Complete logic diagram of an IC containing three OR gates. The Boolean equation describes its operation, telling you that the output (pin 3) is the sum of two OR gates' inputs and a third is included, as indicated by the + sign.

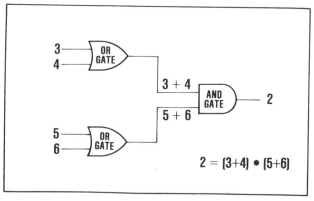

Figure 2-6:   Logic diagram of an IC containing two OR gates and one AND gate.

Next, let's try a more complicated **IC**, the Signetics 10121, 4-wide 3, 3, 3, 3 input **OR-AND/OR-AND**-inverter gate. Figure 2-7 shows a logic diagram for the 10121 **IC**.

If you will examine the logic diagram, you'll see that pin 10 is common to two of the gate inputs (pins 9 and 11). This function is particularly useful if you are working with data control or multiplexing circuits. At this point, there are two ways we can go: 1) Refer to the manufacturer's data book, sales catalog, or information sheets, and see if they give the equations for the **IC** or; 2), refer to the logic

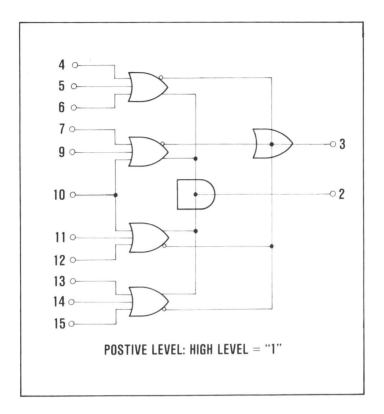

POSTIVE LEVEL: HIGH LEVEL = "1"

Figure 2-7: Logic diagram for a 10121 4-wide 3, 3, 3, 3 input OR-AND/OR-AND-inverter gate

diagram and derive the equation from it. Either way, using positive logic, you should end up with the following equations.

$$2 = (4+5+6) \bullet (7+9+10) \bullet (10+11+12) \bullet (13+14+15)$$
$$3 = \overline{(4+5+6)} + \overline{(7+9+10)} + \overline{(10+11+12)} + \overline{(13+14+15)}$$
$$= \overline{(4+5+6)} \bullet \overline{(7+9+10)} \bullet \overline{(10+11+12)} \bullet \overline{(13+14+15)}$$

Looking at the logic diagram, you can see that there are three inputs to each **OR** GATE (4, 5, 6, and so on) and the outputs of the **OR** gates are fed to an **AND** gate. You should see the same information when reading the equation. We can now ask, "What are the solid bars shown in the equation telling us?" To answer this question, again refer to the logic diagram. You'll notice that the diagram shows a small circle on the **OR** gate outputs. This is an inverting symbol (if logic 1 level is applied to an input, the output level is reversed), the solid bar over $3 = (\overline{4 + 5 + 6})$ etc., denotes inversion. Looking at it another way, the symbol for negation is a *superscript* bar.

Understanding and Using a Logic Diagram

The key to understanding logic diagrams is two-fold: being able to work with Boolean algebra formulas, as we have just explained, and understanding truth tables. Figure 2-8 shows the logic symbols and pin configuration for the quad 2-input positive **AND** gate (7408). If you refer to a manufacturer's data sheet, you'll find the electrical characteristics for this **IC** are listed as shown (this is only a partial list):

PARAMETER

$V_{in}(1)$ Logical 1 input voltage required at both input terminals to ensure logical 1 level at output.

$V_{in}(0)$ Logical 0 input voltage required at either input terminal to ensure logical 0 level at output.

$V_{out}(1)$ Logical 1 output voltage.

$V_{out}(0)$ Logical 0 output voltage.

Now, let's see if we can express the electrical characteristics of the 7408 **IC** in tabular form. Because all four **AND** gates work exactly the same, we only need to examine one to find out how each will perform. For instance, the **AND** gate in the lower left-hand corner, with input pins 1 and 2, output pin 3, will do. Now, if we place the electrical characteristics, given by the manufacturer, in tabular form, the end result is as shown in Table 2-5.

Looking at the truth table for the **AND** gate shown in Table 2-5, we can derive a Boolean algebra formula for the device. Notice it takes a logic level 1 on both inputs to produce a logic level 1 on the output. In other words, 1 and $1 = 1$ or, $A \bullet B = C$, in Boolean algebra.

Dual **J-K** flip-flops such as the 7473 are not expensive, so let's examine the truth table for one of these **IC**'s and see what it can do for us. First, the pin configuration for this **IC** is shown in Figure 2-9.

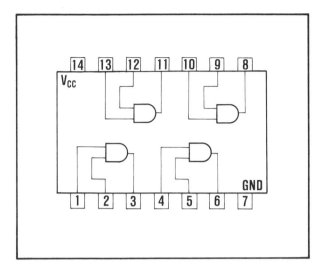

Figure 2-8: Pin configuration and logic symbols for the four AND gates contained in a 7408 IC

INPUT		OUTPUT
PIN 1	PIN 2	PIN 3
1	1	1
0	1	0
1	0	0
0	0	0

Table 2-5: Table showing all the possible input/output logic levels of an AND gate in a 7408 IC. This is called a *truth table*

As you can see by looking at the pin configurations of this **IC**, we are still at a loss without something like an application note to tell us what the **IC** can and will do. This flip-flop is based on the master-slave principle which, as you will remember, means it has two in-dependent storage stages (flip-flops) that require separate clock pulses for each stage to transfer information from the master stage to the slave stage. The sequence of operation will go like this:

1.   Isolate slave stage from master stage. You do this with the clock pulse.

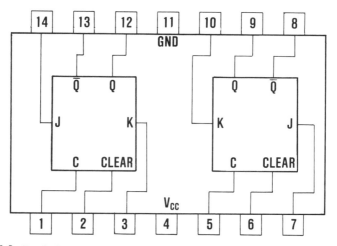

Figure 2-9: Dual J-K master-slave flip-flop (7473) pin configurations

2. Enter information (bit) from **J** and **K** inputs to master stage.

3. Disable **J** and **K** inputs.

4. Transfer information (bit) from master stage to slave stage.

Now we need to know where the information will be in the system (the two flip-flops, master, and slave) after each clock pulse, if we are going to use, or troubleshoot, the **IC**. Where can we get this information? From the truth table for the **IC**. Table 2-6 shows a truth table for the 7473.

Figure 2-10 shows a logic diagram for a 4-bit universal shift register (the 8371). You'll notice the control inputs (load and shift) first pass through buffer amplifiers then into **AND-OR**-Invert gates. These gates are necessary to accomplish a shift-right operation in the register.

Using this design permits either serial or parallel data to be synchronously entered on the falling edge of the clock. This IC has four modes of operation.

1. Will store data.

2. Parallel information will be synchronously entered.

3. Serial data will be shifted to the right.

4. Not true output for the fourth stage (D_0).

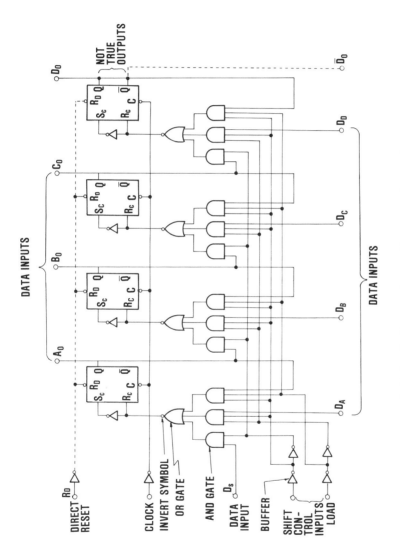

Figure 2-10: Logic diagram for a 4-bit shift register (8371)

EACH FLIP-FLOP		
t_n		t_{n+1}
J	K	$\dot{Q}$
0	0	Q_n
0	1	0
1	0	1
1	1	$\overline{Q}_n$

NOTES:
1. t_n = bit time before clock pulse
2. t_{n+1} = bit time after clock pulse
3. n = time prior to clock
4. Q, $\overline{Q}$ = outputs

Table 2-6: Truth table for a dual J-K master-slave flip-flop (7473)

Table 2-7 shows the effects of load and shift inputs when the data is entered on the falling edge of the clock pulse. After the clock falls, data is tranferred to the outputs. Data can be moved from one register to another by serial shifting. However, a faster method is to use parallel transfer. Many **IC** shift registers (such as this one) include both serial and parallel data entry capabilities.

LOAD	SHIFT	CONTROL STATE
0	0	STORE OR HOLD (WITH CLOCK RUNNING)
1	0	PARALLEL ENTRY
0	1	SHIFT RIGHT
1	1	SHIFT RIGHT

Table 2-7:   Truth table showing the effects of load and shift inputs to a shift register IC (8271)

Working with AND Gates Having
More than Two Inputs

You'll notice that the 7411 IC (see Figure 2-11) contains *positive* AND gates. Let's digress for a moment and ask, what if you encounter a negative AND gate rather than the positive shown in Figure 2-11? The truth table for a negative logic AND gate (we will use a 2-input AND gate for this example) is shown in Table 2-8. Notice that the truth table for the negative logic AND gate is exactly opposite to the positive logic AND gate shown in Tables 2-5 and 2-9.

According to the truth table for a positive logic AND gate (shown in Table 2-9), if we apply a logic 0 and a variable (either 1 or

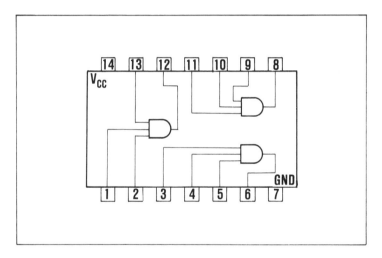

Figure 2-11: Triple 3-input positive AND gate IC (7411)

INPUT		OUTPUT
B	C	A
1	1	1
0	1	1
1	0	1
0	0	0

Table 2-8: Truth table for a 2-input negative logic AND gate

INPUTS		OUTPUT
1	2	3
0	0	0
0	1	0
1	0	0
1	1	1

Table 2-9: Truth table for a positive logic **AND** gate.

0) to the input of an **AND** gate, the output will be 0. But not all **AND** gates use only two inputs. For example, Figure 2-11 shows a triple 3-input positive **AND** gate **IC** (7411).

What are the rules for **AND** gates having more than two inputs? Using one of the 3-input positive **AND** gates shown in Figure 2-11 as an example, the rules can be stated basically the same as shown in the truth table in Table 2-9 (incidentally, these same rules can be used for an **AND** gate with any number of inputs). In other words,

INPUT-PINS 1, 2, 3			OUTPUT PIN 12
0	0	0	0
0	0	1	0
0	1	0	0
0	1	1	0
1	0	0	0
1	0	1	0
1	1	0	0
1	1	1	1

Table 2-10:   Truth table for a single 3-input positive AND gate (7411)

a logical 1 input voltage is required at all three input terminals of a single gate to ensure logical 1 level at the output. On the other hand, logical 0 input voltage on any input terminal will result in logical 0 level on the output. A truth table would be as shown in Table 2-10.

Practical Guide to OR Gates

The pin configuration and logic symbols for the 7432 quadruple 2-input positive **OR** gates are shown in Figure 2-12. The 7432 provides four 2-input or logic functions. Each gate may be used individually or connected serially to provide an equivalent 5-input **OR** function.

The **OR** gate is somewhat different from an **AND** gate. For example, if you inject an input using a logic level 1 to one of the gates shown in Figure 2-12, the output will be a logic level 1. The formula for the **OR** gate states that input pins 1 or 2, 4 or 5, 13 or 12, 10 or 9, equal an output. In Boolean algebra: A + B = C. The symbol (+) stands for **OR** in Boolean algebra. The truth table for a single **OR** gate (positive logic) is shown in Table 2-11.

Guidelines for Using NAND Gates

We have discussed a **NAND** gate in Chapter 1; however, because a **NAND** gate is one of the most often used gates in digital

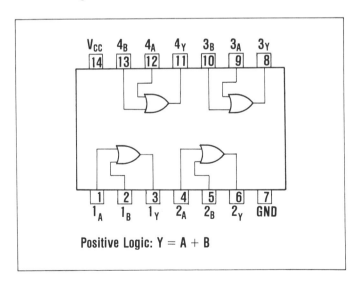

Figure 2-12: Pin configuration and logic symbols for a quadruple 2-input positive OR gate IC (7432)

INPUT		OUTPUT
PIN 1	PIN 2	PIN 3
0	0	0
0	1	1
1	0	1
1	1	1

Table 2-11: Truth table for an individual OR gate in the IC shown in Figure 2-12

circuitry, it's a good idea to understand how it works from all angles. First, let's look again at the quad 2-input **NAND** gates in a 7400 **IC**. These are shown in Figure 2-13.

Referring to a single **NAND** gate logic symbol in Figure 2-13, you'll notice that a **NAND** gate is actually an **AND** gate followed by an inverter. The truth table for any one of the **NAND** gates in Figure 2-13 is shown in Table 2-12.

The operation of a **NAND** gate is represented by the equation $A = \overline{B \bullet C}$ and is read as output A is the result of B and C operating at the input of the **NAND** gate but inverted at the output. As we have explained before, the solid bar means inversion.

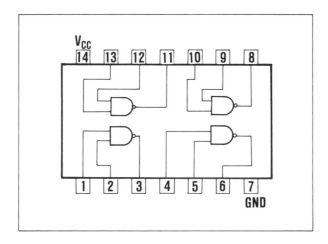

Figure 2-13: Quad 2-input NAND gate IC (7400)

INPUT		OUTPUT
PIN 1	PIN 2	PIN 3
0	0	1
0	1	1
1	0	1
1	1	0

Table 2-12: A single 2-input NAND gate truth table using the first NAND gate pin numbers 1, 2, and 3 in Figure 2-13, as an example

How to Analyze Exclusive OR
Gate Logic Circuits

The **OR** gates shown in Figure 2-12 are called *inclusive* **OR** *gates* because, as the truth table (Table 2-11) shows, any input (or all inputs) that contains a logic level 1 will produce a logic level 1 at the output. Now, what do you do when you need an **OR** gate to produce a logic level 1 when, and only when, one input is a logic level 1? To answer this question, let's first draw a truth table of what we want. Table 2-13 shows the desired operational conditions of the **IC** we would like.

INPUTS		OUTPUT
AT A	AT B	AT C
0	0	0
0	1	1
1	0	1
1	1	0

Table 2-13: Truth table of an IC requirement explained in the text

As you have probably already guessed, it just so happens that an *exclusive* **OR** *gate* has exactly the same truth table as that shown in

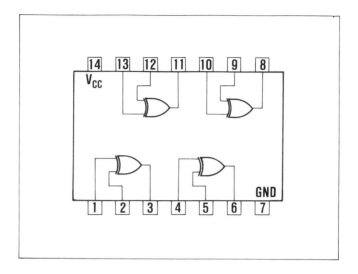

Figure 2-14: Quad 2-input exclusive OR gate

Table 2-13. Now, all we need is the pin configuration and identifying number of an exclusive **OR** gate **IC**. Figure 2-14 shows a quad 2-input exclusive **OR** gate 7486 **IC**. You'll notice when referring to the truth table that you must place a 1 on either input A or B (but not both) in order to have a 1 on the output. To say essentially the same thing using a formula, $A\overline{B} + \overline{A}B = C$.

Understanding the Operation
of a NOR Gate

You have studied the operation of a logic **OR** gate in a previous section. Now, let's see what happens when an **OR** logic gate is followed by an inverter. Figure 2-15 shows a low-cost quad 2-input positive **NOR** gate **IC** (7402) that is a combination of **OR** gates followed by inverters.

When both inputs (any one of the single gates) are logic level 0, the output will be logic level 1. If either of the inputs is logic level 1, you should expect to find the output at a logic level 0. The operation of one **NOR** gate is represented in the truth table shown in Table 2-14.

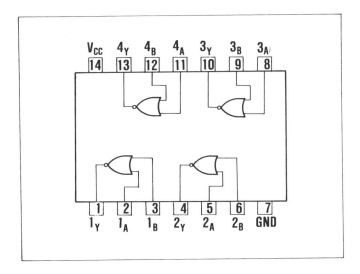

Figure 2-15: Quad 2-input positive NOR gate IC (7402)

INPUT		OUTPUT
A	B	$Y = \overline{A + B}$
0	0	1
0	1	0
1	0	0
1	1	0

Table 2-14: Truth table for the operation of a single positive NOR gate. See Figure 2-15 for the logic symbols and IC package of a 7402 NOR gate

Checking Inputs and Outputs of the RS and D Type Flip-Flops

All flip-flops have two stable states. You can quickly check the states by using a logic pulser and a logic probe, a voltmeter, and/or an oscilloscope. Whatever the state is (a logic high 1 or logic low 0, assuming positive logic), it should remain in that condition until the state is changed by an external signal. Now, all you need to check a flip-flop such as the **RS** (or any other type), are the logic symbol and truth table for the device. Figure 2-16 shows the logic symbol for the **RS** (set-reset) flip-flop and Table 2-15 shows a truth table.

Referring to Figure 2-16, if you apply a logic level 1 to the S input, it should make the Q output go to logic level 1 and the $\overline{Q}$ output go to logic level 0. On the other hand, if you apply a logic level 1 to the **R** input, you should see the output levels reverse. *Note:* You must hold the input you're not using at logic level 0 to produce the conditions listed.

This same information is much better presented in a truth table for the **RS** flip-flop. In Table 2-15, you'll notice that an input of

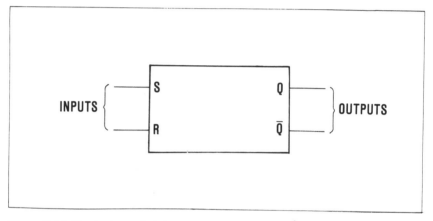

Figure 2-16: The logic symbol for an RS (set-reset) flip-flop

INPUTS		OUTPUTS	
R	S	Q	$\overline{Q}$
1	0	0	1
0	1	1	0
0	0	UNCHANGED	
1	1	NOT PERMITTED	

Table 2-15: Truth table for an RS flip-flop. The unused input must be held at logic level 0 when checking the flip-flop

logic level 1 on both inputs is not permitted. This is because the device will start to race on the output or produce outputs not shown in the truth table. When working with **RS** type flip-flops, you'll get a stable operation only when **S** input is logic level 1 and **R** is logic level 0 (*the flip-flop is reset*), and when **S** is logic level 0 and **R** is logic level 1 (*the flip-flop is set*). Of course, as you can see, logic level 0 on both inputs will leave the **IC**, etc., unchanged.

Once you know how to check an **RS** type flip-flop, the delay (**D**) type is easy because it is much the same. The major difference is that you must go from logic level 0 to logic level 1 on the T input to make the flip-flop store information or to make the device change state (toggle). The basic delay, or **D**-type flip-flop, logic symbol is shown in Figure 2-17.

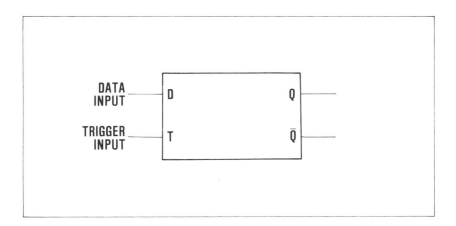

Figure 2-17: Basic logic diagram for a delay or D-type flip-flop

Notice, the outputs are the same as for the **RS** type flip-flop shown in Figure 2-16. However, the lettered inputs are slightly different, which indicates a slightly different type operation. When you are working with the **D**-type, you should expect to see an output that is a function of the input that appeared on the clock pulse (the input on the T input) earlier. Figure 2-18 shows an operational table for a **D**-type flip-flop and an example timing diagram. In order to turn on the **D**-type flip-flop where Q = 1, a 1 must be present at the **D** input and the trigger (clock) input at the same time, as you can see.

The **D** line shows a series of data pulses reflecting a changing data state on the input **D** line; the T line shows clock pulses, and the

Q line shows the resulting output. An example of a low-priced **IC** is the 7474 dual **D**-type edge-triggered flip-flop shown in Figure 2-19.

INPUT		OUTPUT	
D	T	Q	Q̄
0	0	PREVIOUS STATE	
0	1	0	1
1	0	PREVIOUS STATE	
1	1	1	0

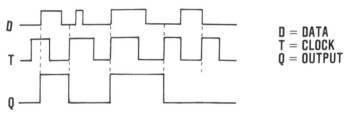

D = DATA
T = CLOCK
Q = OUTPUT

Figure 2-18: D-type flip-flop operational table and example timing diagram

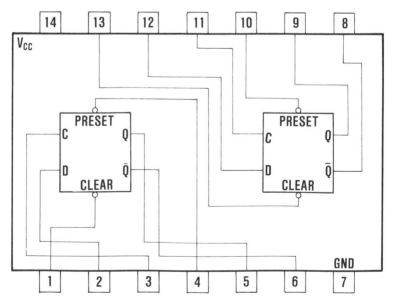

Figure 2-19: Dual D-type edge-triggered flip-flop IC (7474)

CHAPTER 3

Complete Guide to Testing
Digital Circuits

In this chapter you get complete step-by-step instructions for building, testing, and using three of the most important pieces of digital test gear; a logic memory probe, a logic pulser, and a logic monitor. These timesavers will be welcome additions to your workshop — and if you're thinking of starting a home service shop, they can give you the extra edge that's so important in the constantly changing electronics field.

You'll also find other digital troubleshooting equipment is included. For each unit, we will try to outline when and where it should be used, noting limitations. Typical problems will show various methods of obtaining the data required to pinpoint a fault in **TTL, CMOS,** and **MOS** devices.

How to bring yourself up-to-date with actual digital troubleshooting information you can use on the job is one of the most frequently asked questions by electronics technicians today. This chapter answers that question and will help you ride the crest of success to more responsibility and more frequent promotions.

Using a Multimeter for
Digital Circuit Testing

Testing digital integrated circuits requires you to adopt a new workbench philosophy, if all your testing has been restricted to analog circuits. Traditionally, when the individual components of a circuit were accessible, you could use signal generators, diode testers, transistor testers, oscilloscopes and the like to test almost anything electronic. Not today . . . especially when testing digital integrated circuits. In fact, about the only traditional test gear that can be used when working with digital circuits are the multimeter and oscilloscope.

With the multimeter, you can check the power supply voltage of a malfunctioning digital circuit, and other so-called *DC characteristics*. As a practical example, let's assume that you want to check the DC characteristics of the familiar 7400 **IC** . You'll remember that this **IC** is a quadruple 2-input positive **NAND** gate. This gate, like all digital circuits, is simply an ON-OFF switch. ON is represented by a logic level 1 voltage, and OFF by a logic level 0 voltage. The most important question at this point is, "What voltage would you actually measure when a logic level 1 is measured at a certain pin?" To answer questions like this one, we must refer to the **IC** specifications sheet and study the **IC**'s electrical characteristics. Table 3-1 shows the electrical characteristics of a 7400 **IC** gate.

You'll notice (Table 3-1) that some of the parameters are gate input currents (I_{in}). In general, gate inputs, except for those where **CMOS** devices are included, require a current supply, or deliver current to a driver stage or other circuit. If the circuits you are working with involve several gates fanned in or fanned out, it's especially important that you take the current parameters into consideration. Another current parameter that you need to watch is *short-circuit output current* (I_{os}). This parameter is an indication of the ability of the **IC** to withstand a short circuit. *Note:* As a general rule, you should not short more than one internal **IC** circuit during any test.

In some cases, the minimums are the most significant values and, in others, the maximums. The other value, typical (TYP), should not be used as anything except to get an idea how the device should perform during a test run under the specified conditions. To make a logic 0 output, V_{in} (1), test of a gate in the 7400, you can use the test setup shown in Figure 3-1.

PARAMETER	TEST CONDITIONS*	MIN	TYP**	MAX	UNIT
$V_{in(1)}$ Logical 1 input voltage required at both input terminals to ensure logical 0 level at output	V_{CC} = MIN	2			V
$V_{in(0)}$ Logical 0 input voltage required at either input terminal to ensure logical 1 level at output	V_{CC} = MIN			.08	V
$V_{out(1)}$ Logical 1 output voltage	V_{CC} = MIN I_{load} = −400 μA V_{IN} = 0.8V	2.3	3.3		V
$V_{out(0)}$ Logical 0 output voltage	V_{CC} = MIN I_{sink} = 16 mA V_{in} = 2V		0.22	0.4	V
$I_{in(0)}$ Logical 0 level input current (each input)	V_{CC} = MAX V_{in} = 0.4V			−1.6	mA
$I_{in(1)}$ Logical 1 level input current (each input)	V_{CC} = MAX V_{in} = 2.4V V_{CC} = MAX V_{in} = 5.5V			40 1	μA mA
I_{os} Short circuit output current	V_{CC} = MAX V_{in} = 5V	−18		−55	mA
$I_{CC(0)}$ Logical 0 level supply current	V_{CC} = MAX V_{in} = 5V		12	22	mA
$I_{CC(1)}$ Logical 1 level supply current	V_{CC} = MAX V_{in} = 0		4	8	mA

*For conditions shown as MIN or MAX: Minimum = 4.75V, Maximum = 5.25V
**All typical values are at V_{CC} = 5V, TA = 25°C

Table 3-1:   Electrical characteristics of a 7400 IC needed for testing the device (courtesy Signetics)

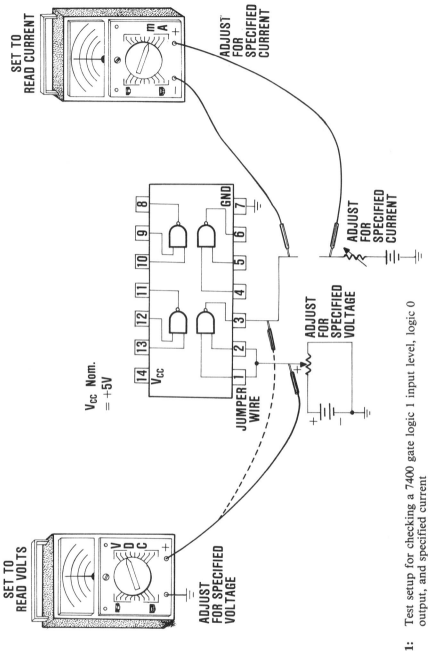

Figure 3-1: Test setup for checking a 7400 gate logic 1 input level, logic 0 output, and specified current

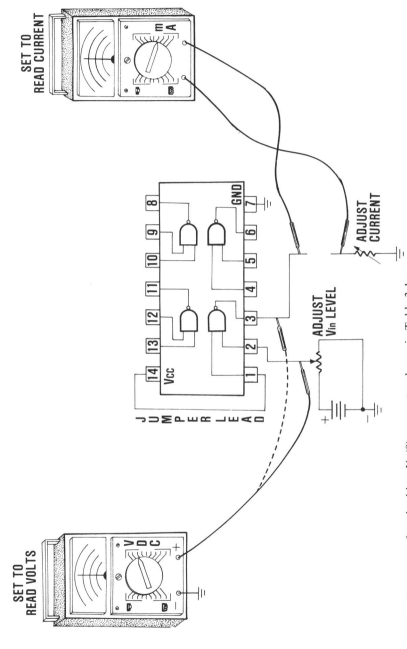

Figure 3-2: Test setup for checking $V_o(0)$ parameter shown in Table 3-1; i.e., to insure logical 1 level at output

The second parameter shown in Table 3-1 can be used to test for the opposite condition; i.e., $V_{in}(0)$, logical 1 level on the output. When you make this test, be sure to remember: *reverse your current meter leads*. Notice, the current meter connections are the exact opposite in Figure 3-1 from those shown in Figure 3-2 for this test. The $I_{CC}(0)$ and $I_{CC}(1)$ typical values given for this IC are about standard for **TTL** logic.

Figure 3-3 shows a test setup that you can use to check the input current under $I_{in}(0)$ conditions. You should check each input pin of the **IC** separately if you are testing the 7400, or any other **TTL** gate, for that matter.

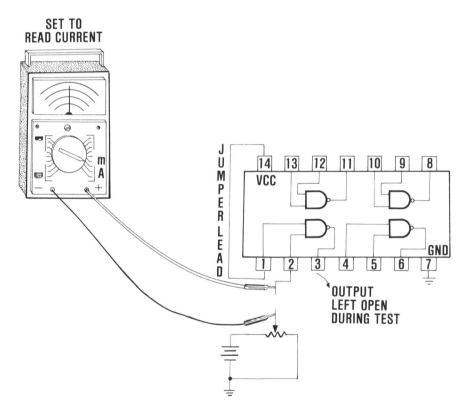

Figure 3-3: $I_{in}(0)$ test setup for each input of a 7400 IC, or similar TTL logic device

You may find that different manufacturers of the 7400 **IC** list lower or higher current values than shown in Table 3-1 for $I_{in}(0)$ and

I_{in} (1), but, in most cases, this is not of major importance. What you are mainly interested in, when testing **TTL** devices such as the 7400, is that each of the inputs has about the same value of input current during testing. If they don't, it's almost certain that you have a bad **IC**.

The next parameter of interest (shown in Table 3-1) that you may wish to check is I_{in} (1). Again, you should check gate input separately. Also, don't forget to check your ammeter for the correct connections, as shown in Figure 3-4. The same rule applies when checking the I_{in} parameter. If all inputs do not have somewhere close to the same current reading, you can suspect a defective **IC**.

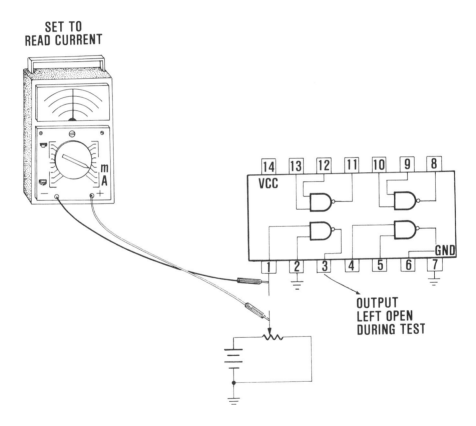

Figure 3-4: Test setup for checking the I_{in} (1) parameter shown in Table 3-1, or any similar TTL device

Referring to Table 3-1, you'll notice I_{os} follows the I_{in} parameters. The test setup for checking the short-circuit output current is shown in Figure 3-5. There are 3 cautions to be observed during this test. These are:

1. *Not more than one output should be shorted at a time.*

2. *A current limited power supply should be used during the test.*

3. *The IC must be disconnected from any original circuits before performing any of these current tests.*

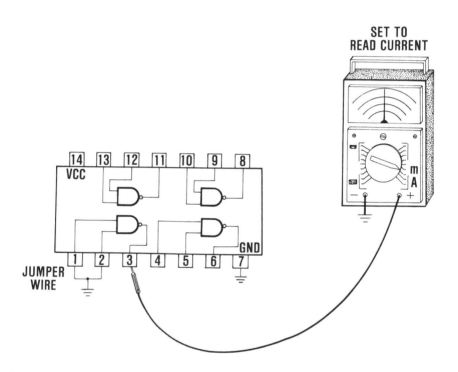

Figure 3-5:   Test setup for checking short circuit output current (I_{os}). *Note: do not make this test without reading cautions listed in text*

The final parameter (I_{CC}) may be tested for logical 0 level supply and logical 1 level supply (all gates to be in the same condition). Sometimes you will find a spec sheet will list the current for each gate. In this case, the total current (I_{CC}) will be the sum of the rated current per gate. See Figure 3-6 for test setup.

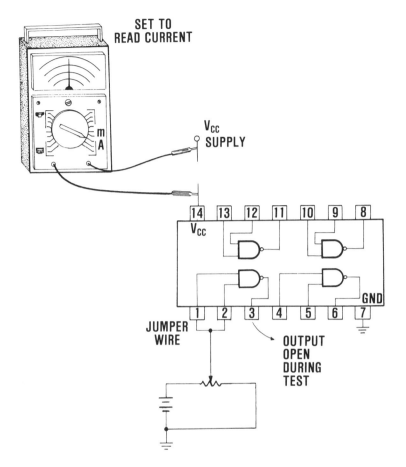

Figure 3-6:   Test setup for checking supply current

How to Test with a Logic Probe

One test instrument that almost anyone who works with digital circuits has become familiar with is the *logic probe*. The instrument is hand-held and has one or more indicator lights near its probe tip. An example is shown in Figure 3-7. The logic probe shown also includes a memory indicator and memory reset switch. The purpose of the memory is to make it possible for a change in logic level to be observed. If you are working with a single indicator light probe, a lit indicator is a logic 1 and an unlit indicator is a logic 0. The probe in Figure 3-7 has two logic indicators; a light for the logic 1 state, and another for the logic 0 state.

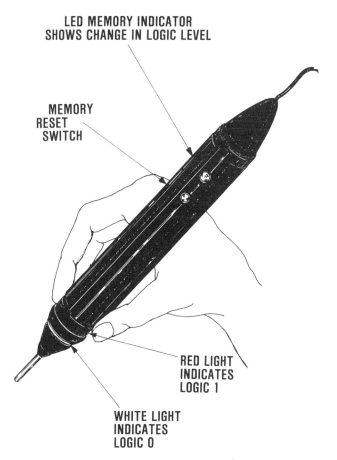

LED MEMORY INDICATOR
SHOWS CHANGE IN LOGIC LEVEL

MEMORY
RESET
SWITCH

RED LIGHT
INDICATES
LOGIC 1

WHITE LIGHT
INDICATES
LOGIC 0

Figure 3-7: A probe with two logic indicator lights near its tip. From *Practical Handbook of Solid State Troubleshooting*, page 180.

To use a logic probe, connect the two wires that are provided for an external connection to the power supply (generally, V_{cc} and ground; the red wire is positive and the black is negative). When testing with the probe, simply touch a PC card run or the pin of an **IC** and observe the indicators, or indicator, whichever the case may be.

Because a probe works best when observing low-repetition rate pulses of short duration, you should use the probe in either of the following two ways. One, you can start the circuit under test operating at its normal clock rate and then check only key signal lines or, two, if the clock rate is too fast to be observed effectively, slow the system clock rate down by using a low-frequency clock source until you can observe the state using the probe.

I would like to say, "That's all there is to it." Unfortunately, that *isn't* all there is to it because, when you slow the clock down, you'll probably find that the circuit, due to propagation delay, no longer reacts the same. Many prototype problems that occur are due to the delaying of parallel signals, extraneous voltage pulses produced by propagation delays in two or more logic paths, or because of PC board inductance and capacitance (you'll remember that both of these are frequency sensitive). If you run into these problems, it's possible that you may have to use an instrument such as a logic comparator or analyzer rather than a logic probe (more about these instruments in the following pages).

Testing instructions, particularly those for digital circuits, may seem very loose if you are familiar with electronic component specs in general. When checking logic levels, it's important to remember that *absolute amplitudes are unimportant.* As has been explained previously, logical states are defined by threshold levels of the logic family you're checking. For example, for **TTL** the low threshold is 0.4 volts and the high threshold is 2.4 volts. When the amplitude of the signal you are checking is *less* than 0.4 volts, it is considered to be at logic 0 level and the logic probe light that indicates logic 0 should light (see Figure 3-7). When the **TTL** signal under observation is above 2.4 volts, it is at a logic 1 level and the upper light at the probe tip should light. However, some probes will instantly recognize high, low, or intermediate levels, open circuits and pulsing modes.

In summary, the low-cost logic probe works best at low-frequency levels but can be used on high-frequency circuits, if you restrict your tests to checking the key control lines of the system. (Incidentally, there are high-speed logic probes that capture pulses as short as 10 ns, but they cost a bit more.) Also, if your logic probe has a memory, you aren't left defenseless when trying to cope with 50 ns pulses, which are more than long enough to trigger **IC** logic devices. This can be a problem if you do not have a high-speed triggered sweep oscilloscope. A logic probe can overcome the oscilloscope problem.

How to Recognize IC Failure Modes

A solid grasp of the contents of this section will help you in several ways. It will make your job easier, speed up your work, and may increase your earning power. Furthermore, most of your prototypes will have a few design/fabrication bugs, and any system can malfunction due to a defective component. Unfortunately, troubles like these are generally hard to identify and track down. Unlike many other solid state circuits, you must know what the output of a digital circuit will be for a given set of inputs. Therefore, when you are troubleshooting, you may have to apply different sets of inputs in the form of 0's and 1's before you can locate a faulty component or completely test a unit.

Basically, you'll encounter only five kinds of failure when working with the various logic families (**MOS, CMOS, TTL, PAL,** etc.). These are:

1.   Defective internal logic.
2.   Input/output shorted to ground.
3.   Input/output shorted to V_{CC}.
4.   Input/output open.
5.   A short between the **IC** pins.

The ways in which a failure occurs in the various logic families listed are similar, although some logic families may be more prone to one particular type of failure. For example, a certain **IC** may be more likely to fail due to defective internal logic because of poor manufacturing practice. Or, you may have more trouble with shorting between pins, due to poor construction techniques. However, whatever the cause, the trouble symptoms can be identified as follows:

The first failure, that of defective internal logic, will result in erratic readings when you measure the circuit parameters. For example, you may read 0 logic on the output when it's supposed to be a logic 1, or vice versa. Of course, an erratic output such as this will probably bring all operations to a halt if the **IC** is operating in some type of equipment. To see why you will experience erratic readings, let's examine the failure of components within the input circuitry of a single gate of a 2-input **TTL NAND** gate such as shown in Figure 3-8.

Let's say that Q_1, the input transistor, opens. Transistor Q_2 will then start conducting and immediately drive the bottom totem-pole transistor, Q_4, into conduction. This will cause the output to go to a logical 0.

Next, let's assume Q_2 develops an open. In this case, it will cause transistor Q_3 to go into conduction, with the end result a steady

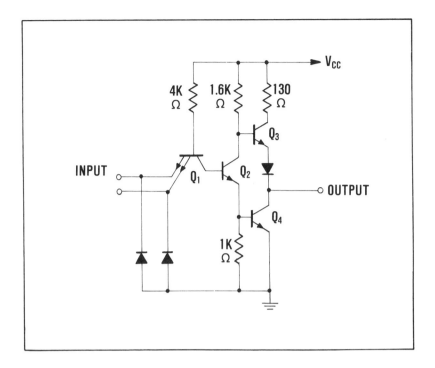

Figure 3-8: Input circuitry for a single gate of a 2-input TTL NAND gate IC

state 1 on the output. What this comes down to is that you never know what an internal failure will do. It all depends on what happens inside the **IC**.

The second and third failures, those of input/output shorted to ground or V_{CC} , are also tricky because, in some cases, normal circuit operation continues with only occasional circuit malfunctions. In most cases, however, some expected system operation will come to a halt and you know you have a trouble in that area, which makes it fairly easy to find. In general, with a short to ground all the affected circuits will remain in a logic 0 state. If the short occurs in a V_{CC} circuit, all affected stages will remain permanently in a logic 1 state.

The fourth failure, open input/output, will affect different circuits, depending upon whether it occurs at an input or an output. For example, an open input will probably cause the input to float at some unpredictable level and generally will have no effect on the input signal. On the other hand, an open output will usually cause the **TTL** device to go to a high logic level somewhere near 1.5 volts for both **TTL** and **DTL**. Although this is slightly less than what is correct for a logic 1, you will probably find that all following inputs will act as if they are at a high level. A rule to remember is: In **TTL**, an open (or floating) input is interpreted by the **IC** as a logic 1 level.

The fifth failure, a short between two pins, can cause you to want to quit the business. The problem is that when two pins of an **IC** are shorted together, it causes the outputs of the **IC**'s (which drive those two pins) to attempt to pull the other **IC** pins to logic 1 or 0. As with any short, the circuit will draw excessive current. In this case, it results in a logic 0 level on both outputs of the driver **IC** stages. Now, it's possible that when both **IC** outputs are working together (both high or low level), your test will show the circuit operating properly. But when they are driven into alternate states (one output 0 and the other 1), you will not measure anything but a low state on the output. There is a good chance that the circuit, due to the burned out components, will finally come to a complete stop. Assuming you spot the trouble before excessive heat destroys some component, the best way to troubleshoot the problem is to have one driving output at logic 1 and the other at logic 0, then check the logic level of the input pins to the driven **IC**.

EXPERIMENT 3-1

Digital Circuit Testing

Let's look at an example circuit and examine its operations. The circuit is shown in Figure 3-9 and is the master-slave **J-K** flip-flop that we have discussed before. But, this flip-flop is constructed using two 7400 quad 2-input **NAND** gates and one 7407 hex inverter buffer/driver, all of which you are now familiar with.

If you would like to construct the circuit shown in Figure 3-9 and run some experiments, you'll find the **IC**'s are all very inexpensive. Switches S_1 and S_2 (on the **J** and **K** inputs) are single-pole-single-throw (SPST) toggle switches, readily available in almost any electronics supply store. Figure 3-10 is an illustration showing toggle, and push-button switches.

Switch 3 should be a snap-acting push-button switch. This switch will produce a clock waveform when you press the switch button and is *debounced* by the 7407 **IC** and its associated circuit. Using a manual switch trigger pulse will most likely result in oscillations and

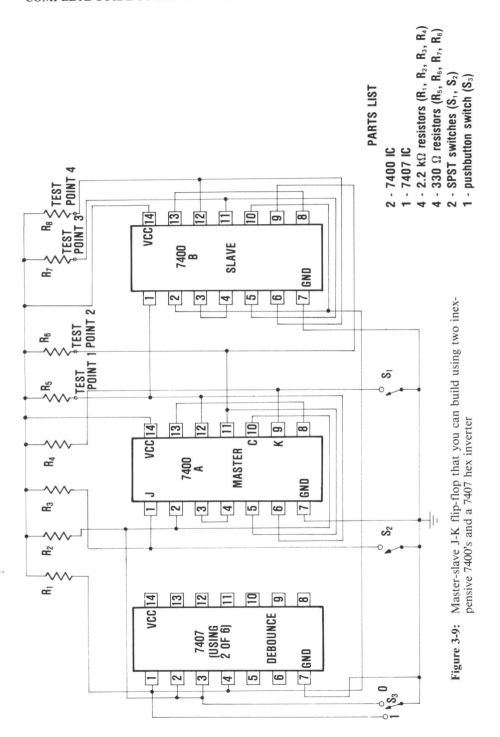

Figure 3-9: Master-slave J-K flip-flop that you can build using two inexpensive 7400's and a 7407 hex inverter

PARTS LIST

2 - 7400 IC
1 - 7407 IC
4 - 2.2 kΩ resistors (R_1, R_2, R_3, R_4)
4 - 330 Ω resistors (R_5, R_6, R_7, R_8)
2 - SPST switches (S_1, S_2)
1 - pushbutton switch (S_3)

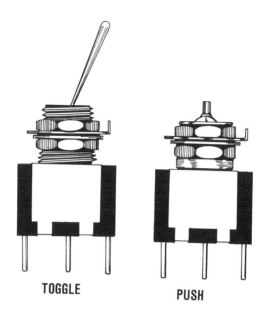

TOGGLE   PUSH

Figure 3-10: An example of a toggle switch and snap-acting push-button switch that can be used during construction of the circuit shown in Figure 3-9

indeterminate triggering of the master section **IC**, is you do not include the clock debouncing circuit. The *master section* is made up using the 7400 **IC** labeled (A). The *slave section* is made up using the 7400 **IC** labeled (B).

After the wiring is complete (*leave the power off until the wiring is completed*), you can proceed with the testing. You'll need a voltmeter and clip leads to connect to the test points shown in Figure 3-9 or, if you have them available, you can connect LED's in series with resistors R_5, R_6, R_7, and R_8 (break the line at the test points to insert the LED's).

The next thing you'll need to perform the tests is an appropriate truth table. The one shown in Table 3-2 can be used to set up the input switches *before the clock pulse is initiated.*

First, do not touch the clock switch. Set switch S_1 to ON (1), and switch S_2 to OFF (0). You should not see any output at the slave section. The master section is in its memory mode at this time. Now, depress the clock switch button (S_3). You should see an instantaneous

S_1	S_2	S_3	Tp_3	Tp_4	MODE
0	0	P	Q_{t-1}	$\overline{Q_{t-1}}$	MEMORY
1	0	P	1	0	S_1 SET
0	1	P	0	1	S_2 RESET
1	1	P	1	1	INVALID

P = S_3 PRESSED FOR COMPLETE CLOCK PULSE WAVEFORM

Table 3-2: Truth Table to be used for testing the master-slave flip-flop circuit shown in Figure 3-9

reading of logic level 1 on test point 2 and logic level 0 on test point 1. Without releasing the switch, check test points 3 and 4. There should be no change in state. You have just placed the slave section in its memory mode.

Next, release the clock switch button. This action should cause test point 3 to go to logic level 1 and test point 4 to go to logic level 0. This step gated the slave section which, in turn, produced these outputs and completed the cycle; i.e., stepped the data through the master-slave sections.

In summary, what you should have learned from this testing experiment is: The master section of a master-slave flip-flop responds to the **J-K** inputs only while the clock input is at logic 1. The slave section will only respond when the clock input goes back to logic 0. We could have used an inexpensive master-slave **IC** such as the 7473 dual **J-K** for this experiment. However, you would not have had access to the ouputs of the master sections. Using the circuit we did, allowed you to see each action step-by-step.

How to Use a Digital Pulser

A digital pulser is frequently called a *logic pulser* but, whatever it is called, it's simply a pulse generator that usually provides either a single-pulse and/or a continuous-pulse stream. In passing, we should mention that, although a pulser is not absolutely essential to some troubleshooting jobs, it is a valuable troubleshooting tool. In general, most conventional pulsers provide somewhere near 700 mA of current, which is all you'll need to force most **IC**'s to change states. Some commercial models are pencil-size and will pulse any family of digital circuits. To use a digital pulser, first connect the ground clip to ground and then place the tip of the pulser to the input of the suspect digital **IC**, etc. Figure 3-11 gives you a general idea of the pulser placement.

You will generally find that a pulser has a trigger button. When you press it, you should see the logic element you are testing change state (some digital logic testers emit an audible tone; high tone for the high logic and low tone for the low logic state). There are military pulsers on the surplus market that are designed for various uses. But, if you use one of these, remember, it is important that the

Figure 3-11: Pencil-size digital pulser being used to pulse a digital element

pulser you choose has a very narrow pulse to prevent any **IC** damage that might be caused by applying too much current. Furthermore, like all test equipment, to prevent circuit loading, the pulser should have a high impedance input.

You can signal-trace with a logic pulser and logic probe by simply following the pulser's signal with the probe. For example, if you find current flowing from one gate to another, but there are no pulses, then either an input or an output is shorted. The symptom you'll first notice in a situation like this is that the pulser will have no effect when you touch the tip to the shorted circuit because, when there is a short to ground, even a pulser cannot change the state of the circuit under test. *Note:* Remember, a signal line may be at logic 1 or logic 0 and be correct, depending on whether you are checking positive or negative logic. In other words, a certain circuit may be shorted to a logic high.

As an example of how you can use a pulser and logic probe, let's say you want to check a single **NAND** gate, as shown in Figure 3-2. Referring to this figure, you'll notice that when you are testing each gate, one of the inputs must be tied to a proper input level (in this case we used V_{cc}, which is a high). To test the same gate using a pulser and logic probe, your connections would be as shown in Figure 3-12.

The pulser is placed at pin 2, while the probe is placed at the output, pin 3. You should read a logic level 1 on the probe, since the pulser is normally low until you press the trigger key. Pressing the logic pulser trigger switch should cause the probe to read a change of state.

Observing Signals with a Logic Monitor

A logic monitor (also called a *logic clip*) automatically displays static and dynamic logic states of digital **IC**'s. Most of them work with **DTL, HTL, TTL,** and **CMOS** dual-in-line package **(DIP)** **IC**'s. Typically, they have 16 LED displays. One such instrument is shown in Figure 3-13.

The monitor is simply piggybacked onto either a 14 or 16 pin **DIP IC**. It draws power from the circuit under test. If you have made good contact between all monitor clips and **IC** pins, the monitor LED's will indicate the logic levels (when an LED is ON, it indicates logic level 1 and OFF is logic 0). Basically, the monitor is nothing but

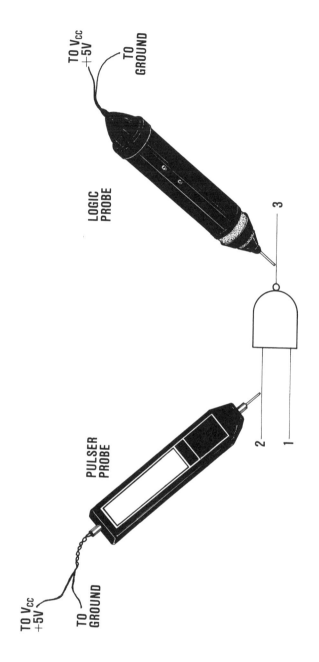

Figure 3-12: Testing a NAND gate with a logic pulser and logic probe. The gate symbol is taken from Figure 3-2

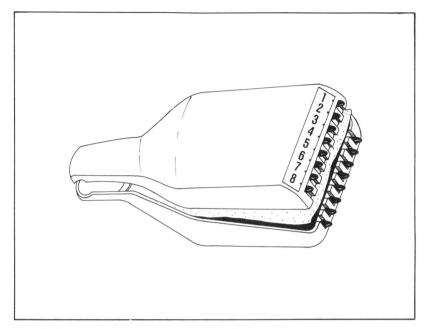

Figure 3-13: Logic monitor that can check all pins of a DIP digital IC. From *Practical Handbook of Solid State Troubleshooting*, page 180.

14 or 16 logic probes all connected and functioning together. However, there are no special external power connections, as with a logic probe.

The logic monitor, like the logic probe, is best suited for low-frequency operation. Therefore, about the same problems are encountered when using the monitor as when using a probe. For example, you may have to try and slow some clock signal down by using a low-frequency clock source. But, as was explained in the previous section about using a probe, due to propagation delay, slowing a clock down can cause problems.

Using Your Oscilloscope to
Test Digital Circuits

An oscilloscope can find much use in digital testing, especially if you have a dual-trace scope with triggered sweep. However, you'll generally find the older scopes unsatisfactory because neither time nor frequency measurements can be made and, in most cases, the vertical axis is not calibrated, preventing you from making threshold measurements.

When using a scope to check digital pulses, it is very important you realize that just because you see a good-looking digital signal on the viewing screen, it does not necessarily mean the circuit is performing properly. To understand why this statement is true, let's review the threshold levels of the **TTL** logic family. The low threshold is 0.4 volts and the high threshold is 2.4 volts. If you measure a pulse *amplitude* of *less* than 0.4 volts, it is a valid logic 0 level. Or, if you measure a pulse *amplitude* of *more* than 2.4 volts, it is a valid logic 1 level. Now, we can again say that when you are viewing logic levels, absolute amplitudes are unimportant. In fact, when using an oscilloscope, you must check to see if the data is a *valid logic level.*

The scope is excellent for checking pulses for distortion. When you are viewing a pulse or pulse train, look for preshoot, overshoot, delay in rise or fall time, too long or short a pulse width, and a delay between input pulses to a logic circuit, any one of which could cause a trouble in the circuit you are troubleshooting or building. A dual-trace scope is especially useful when making all these checks. Incidentally, there are oscilloscope accessories you can connect to the input of your scope that will enable you to check several input and output pulse trains at the same time. However, at this time, they are out of the price range of most small shops and home experimenters and, therefore, will not be covered in this book.

Regardless of low-priced scope limitations you can use them, when working with digital equipment, to check some circuits. For example, if the signal you are checking is of a moderately high frequency, a low-cost (narrow bandwidth) scope may perform quite well. When you are attempting to observe a pulsed signal in logic circuits, you'll get better results with a x 10 (low capacitance) probe. Keep the probe ground clip as close to the point of measurement as you can and, by all means, keep the ground lead as short as possible. Another hint: *Do not* use coax for direct connections. Always use the probe. If this seems like strange advice, there is a reason. Coax cable can introduce excessive capacitance into your test setup (the longer the coax lead, the more capacitance), which, in turn, can temporarily (during testing time) disable circuits such as flip-flops, etc. By the way, to play it safe, every time you move the x 10 probe from one test point to another, re-adjust the probe. With a compensated probe, you can encounter considerable error in your readings (when working with high frequencies) if you don't adjust it properly.

Figure 3-14 shows several important pulse characteristics that you can examine with an oscilloscope that has both good *vertical deflection sensitivity* and *vertical bandwidth.*

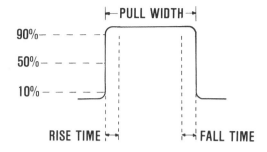

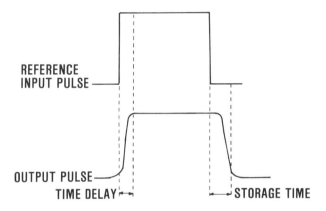

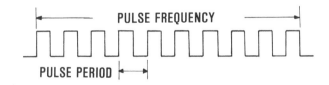

Figure 3-14: Important digital pulse characteristics that can be examined with a good quality oscilloscope

Getting the Most Out of a Current Tracer

Your best solution to a wired-**OR**/wired-**AND** trouble (as wells as shorts and opens when you are working on PC cards) is to use a current tracer. The current tracer operates on the principle that whatever is driving a shorted (low impedance) point must be delivering the greatest amount of the current. Therefore, you should be able to trace the path of the current directly to the low impedance point. A good rule to follow is: If the troubleshooting symptoms indicate an excessive current flow, use of a current tracer is indicated. Figure 3-15 shows the wired-**OR**/wired-**AND** configuration (the connection of two or more open collector or tri-state logic outputs to a common bus, so that any 1 can pull the bus down to 0 level).

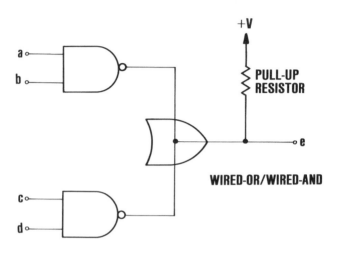

Figure 3-15: Wired-OR/wired-AND configuration

To detect a wired-**OR**/wired-**AND** fault, you can align the mark on a current tracer tip along the length of a printed circuit trace, near the pull-up resistor, and adjust the sensitivity control until the indicator just lights. Then you can move the current tracer along the trace (line), or place it directly on the gate's output pins. Only the malfunctioning gate will cause the indicator to light up.

As you can see from this, although a current tracer looks somewhat like a logic probe and logic pulser, in that it has power leads on one end and a tip on the other, it operates quite differently. The tip does not have to make direct contact with the circuit under test during use. The tip contains a magnetic sensor that is used to monitor the field produced by the current flow in the circuit you are checking. Of course, this makes the instrument excellent for finding low impedance faults. Also, if you are not getting an input signal, you can place a logic pulser at the driving point (pin, etc.). Some of the faults you can locate with a pulser and current tracer are:

1. Shorted inputs of **IC**'s.
2. Bad solder jobs on PC boards (shorted lines due to solder bridges).
3. Shorts between conductors in connections, cables, test leads, and other similar transmission lines.
4. V_{CC}-to-ground shorts.
5. Opens and shorts in a PC card.
6. Signal lines between **IC**'s.
7. When a buffer is driving numerous inputs and one input is shorted to ground.
8. Stuck wired-**AND** gates.

The last item, number 8, can be very difficult to find without a current tracer. Typically, one of the open collector gates connected in

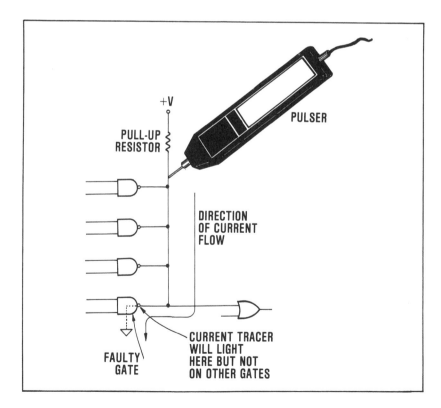

Figure 3-16: Test setup for locating wired-AND troubles

the wired-**AND** configuration may still continue to draw current after it is supposedly non-conducting. Referring to Figure 3-16, you'll see basically what is happening during the trouble and how to locate wired-**AND** problems with a current tracer/logic pulser combination.

PROJECT 3-1:

Building a Logic Memory Probe

The logic memory probe schematic diagram shown in Figure 3-17 can be used to construct a simple but effective instrument that you can use to determine the logic condition of a digital circuit with pulse durations as short as 50 nano seconds.

You will need two **IC**'s for this probe; the 7404 (a hex inverter) and an N8T22A (a retriggerable one-shot multivibrator). The Signetics N8T22A is a direct pin-for-pin replacement for the 9601 retriggerable one-shot. Therefore, either of these **IC**'s will do the job. You will also need a transistor for the input. The one used (a general purpose 2N4401), is fairly inexpensive. The purpose of the transistor is to provide a high input impedance and to work as a buffer for the input to the hex inverter **IC** at pin 11. If you happen to have a spare parts box containing solid state components, you can use any transistor with approximately the same electrical characteristics as the 2N4401. For example, a 2N222, usually found in VHF amplifiers and oscillators, will more than likely work.

Next, the three light-emitting diodes (LED's). In all cases, it is necessary to limit the amount of current through an LED. Resistors R_5, R_6, and R_7 are current-limiting resistors. The 330 ohm value of these resistors will protect the circuit from excessive current flow and provide satisfactory operation of the LED's.

Generally, the continuous forward current in LED's is from 5mA to 40mA, and the forward voltage (V_E) drop of LED's ranges from 1.65 to 2.2 volts. To calculate the approximate value of a current-limiting resistor, you can use the formula:

$$R_L = \frac{V_{CC} - V_F}{I_F}$$

where, R_L = value of current-limiting resistor

V_I = forward voltage drop of LED

I_F = forward current through LED

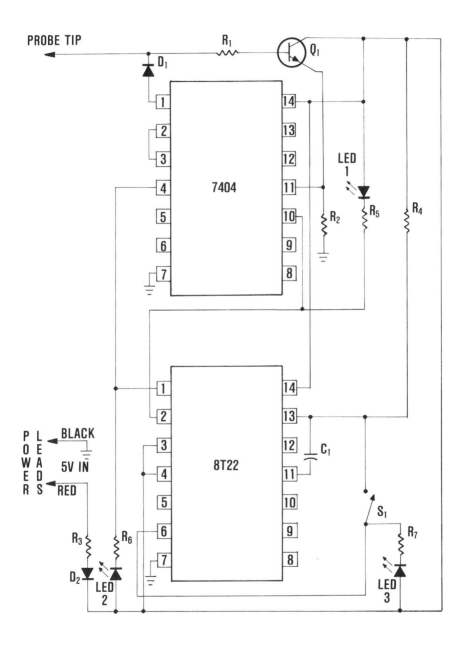

Figure 3-17:   Digital memory probe circuit diagram (parts list on page 94)

PARTS LIST FOR FIGURE 3-17

2 - 1N914, or equivalent, diodes (D_1, D_2)
1 - 7404 IC
1 - 8T22 IC, or equivalent
3 - LED, operating 1.6 to 2.5 volts. 1.5 volts, no light (3 volts will usually cause burn-out)
1 - 22 μF capacitor (C_1). Optional, see text.
1 - 1 kΩ, ¼ watt resistor (R_1)
1 - 390Ω, ¼ watt resistor (R_2)
1 - 6.8Ω, ¼ watt resistor (R_3)
1 - 30 kΩ, ¼ watt resistor (R_4). Optional, see text.
3 - 330Ω, ½ watt resistors (R_5, R_6, R_7)
1 - snap-action pushbutton switch (S_1)
Hook-up wire, insulating spaghetti, solder, etc.

In general, the value of a current-limiting resistor for an LED is not critical. After you have made your calculations, you'll probably have some oddball resistance value. You can vary from your calculated value quite a bit and still have satisfactory operation. For example, a couple of hundred ohms, in many projects, is acceptable although, in most cases, you are better off to use 330 ohms, which is more easily obtained.

For very compact construction, it's best to use a small capacitor for C_1. While they cost a little more, a dipped tantalum (electrolytic) capacitor is probably the smallest, in physical size, you can get. Whatever type capacitor you use, it should have a value of 22 μF and a working voltage of at least 10 volts. A dipped tantalum capacitor made by Panasonic (stock number TAC010), with a value of 22 μF, and rated for 25 volts, costs slightly over a dollar for a single unit.

You can use an external power source to operate the probe, or use alligator clips on the power leads and draw the operating power from the circuit you are testing. The purpose of diode D_2 and the resistor R_3 is to provide "wrong hook-up" and over-voltage protection during use of the probe. The other diode, D_1, acts as a buffer for the transistor input. It protects the transistor from excessively high inputs and also helps maintain a high input impedance. The 1N914 diodes suggested, are fast logic devices with the characteristics listed in Table 3-3. Any diodes with similar operating parameters can be substituted.

DIODE	PEAK REVERSE VOLTS	MAX. FORWARD VOLTAGE AT MAX. mA	MAX. FORWARD mA AT MAX. V	MAX. REVERSE μA
1N914	75	1	75	0.025

Table 3-3: Operating characteristics of diodes (D₁ and D₂) shown in Figure 3-17

Capacitor C_1 and resistor R_4 control the time constant for the 8T22 **IC,** therefore you can choose any two values that will provide a satisfactory operation within the limits of the circuit design. For example, 22 μF and 30 k ohms have a time constant of about one-half second and are what we have used in Figure 3-17.

An rf probe, penlight case, aluminum cigar holder, or any similar tube can be used for a case. The penlight case is probably the best because you can use the built-in switch. However, you will be very restricted in building room if you use a flashlight case. On the other hand, although the other type cases may give slightly more room inside for the circuitry, you will need to mount the switch (S_1) on the outside of the probe case.

The two alligator clips should be color coded, one red and the other black, and have suitable length (about 18 or 20 inches) of stranded power cable. Zip cord (ordinary lamp power cable) will make a satisfactory power lead. Also, small diameter coaxial cable can be used to obtain a more professional looking job.

How to Test the Probe

Step 1: Connect the alligator clips to a variable output DC power supply that is *set to 0 volts out.* Watch the polarity! Black to the negative terminal, red to the positive.

Step 2: Close switch S_1 by pressing the snap-action pushbutton switch. Now slowly adjust the DC voltage supply, starting from 0 volts and advancing toward 5 volts, all the while watching the LED's. At 2.8 volts, you should see the memory LED flash on, then off, as you continue to increase the DC supply output voltage on up through 4.1 volts.

Step 3: Adjust the DC supply until its output is 5 volts.

Step 4: Touch the probe tip to the negative terminal of the power input cable (any ground point). You should see one LED light and another come on for just a short duration of time (about one-half second, with the components shown). The LED that stays on (during the time you are touching the common lead) is the low logic indicator. The LED that flashes on and off (memory LED) glows to indicate a positive/negative-going pulse. When you remove the tip from the negative voltage circuit, you should see the low logic indicator LED go out and the memory LED should glow again for about a half-second.

Step 5: Touch the probe tip to the positive terminal of the DC power supply. Again, you'll see the memory LED stay on for about one-half second and go off. However, you should also see another LED come on and stay on. This LED is the high logic indicator.

Step 6: Your last check is to place switch S_1 into the other position (memory operation). Now touch the probe to either the negative or positive DC power source terminal. You should see the memory LED come on and stay on until you return the push-button switch back to the other mode of operation. In case you are a bit confused at this point, the memory mode (one condition brought on by pushing switch S_1) is designed to stay on, once it is triggered, until you cut it off. The stretch mode (the other position of switch S_1) is designed to observe pulses of short duration that the logic high LED and logic low LED can't catch.

How the Probe Works

If you do not have the probe tip in contact with any circuit (i.e., open) the input to the pin 11 of the 7404 **IC** is at a low logic and the other input, pin 1 of the 7404 IC, is at logic high (assuming you have the power leads connected to a 5 volt DC supply). LED's 1 and 2 are off under these conditions, due to the action of hex inverter connections.

Now, let's say you touch the probe to a logic low (0.8 volts or less). This places pin 1 of the 7404 at a logic level 0 and, due to the wiring of the 7407, transistor Q_1 is cut off. During this time, LED 2 turns on, indicating a logic low level input at the probe tip. When LED 1 turns on (high logic), you'll find transistor Q_1 conducting and LED 2 off. Any sudden change of logic level at the probe tip will produce a negative-going pulse at the input of the 8T22 **IC** (pin 1 or 2), which, in turn, will trigger the multivibrator, resulting in LED 3 turning on for the time constant (set by C_1 and R_4) you have selected.

PROJECT 3-2:
Constructing a Logic Pulser

Figure 3-18 shows a logic pulser schematic that is simple and easy on the pocketbook. The electronics can be housed in a hand-held probe. The pulser is programmed by pushing the switch (S_1) once for a change in logic level output.

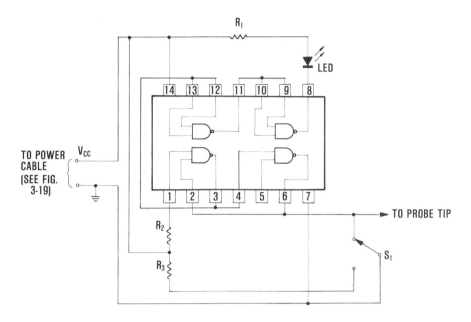

PARTS LIST

1 - 7400 IC
1 - 220 resistor (R_1)
2 - 1.8 k resistors (R_2, R_3)
1 - Pushbutton switch (S_1)
1 - LED
Hook-up wire, spaghetti, solder

Figure 3-18: Logic pulser schematic and parts list. V_{CC} = 5 VDC for nominal operation

A power cable with built-in probe protection should be constructed by using a short length (about 18 inches or so) of flexible cable and two protection diodes, as shown in Figure 3-19.

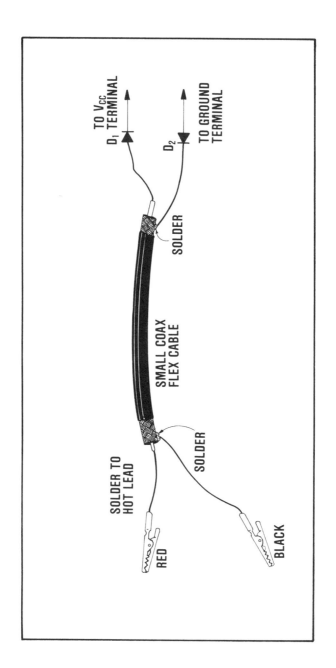

Figure 3-19: Power cable with instrument protection diodes for the logic pulser

How to Test the Pulser

Step 1: Connect the alligator clips to a variable output DC power supply that is *set to 0 volt out.*

Step 2: Connect a DC voltmeter between the probe tip and the supply's negative terminal (ground).

Step 3: Slowly raise the variable DC power supply output voltage to +5 volts. Your voltmeter should read about +0.7 VDC. The pulser is now in a logic low output condition.

Step 4: Press the push-button switch (S₁). Your voltmeter reading should jump to about + 3.6 volts and the LED should come on.

If each of the four steps is performed and each voltmeter reading is fairly close to the ones given, your logic pulser is operating correctly. Incidentally, the procedure given is for testing positive logic. To test negative logic circuits (where a logic low input is required), simply keep the pulser switch pressed down and then release if for short pulses.

PROJECT 3-3:

Building a Logic Monitor

A logic monitor (clip) has one advantage over a logic probe — it can be used to check for correct timing between a number of signals from the same **IC**. The logic monitor described in this project allows *all* of the pins of most **IC**'s to be examined simultaneously, which means you will no longer feel helpless when faced with a timing problem.

Actually, the basic circuit for a logic monitor can be any one of the LED indicator circuits shown in Figure 3-20. However, C and D are particularly good because they will reduce the loading effect on the **IC** under test. Whether you use a bipolar transistor driver, Darlington transistor, or **IC** LED indicator circuit, it is important that you have as high input impedance as practical (1,000,000 ohms or more is best).

When you build your monitor, you first have to decide what number of pins you wish to check. For example, there are 14, 16, and more pin **DIP IC** packages. This is important because you must have 1 LED indicator circuit for *each* pin connection of the **IC** package you want to monitor.

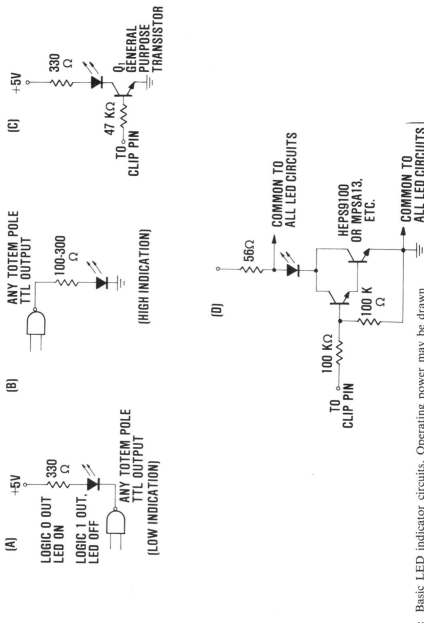

Figure 3-20: Basic LED indicator circuits. Operating power may be drawn from the IC under test.

Darlington Transistor
14-Pin IC Monitor

If you chose to use the Darlington transistor circuit shown in Figure 3-20 (D) and build a 14-pin **DIP** package **IC** monitor, you'll need the components listed in Table 3-4. If you want to build a 16-pin **IC** monitor, simply add two transistors, two LED's, and two each of resistors R_1 and R_2 to the list. Add eight more of each of these components for a 24-pin monitor.

14 DARLINGTON TRANSISTORS, MPS13, GE-64, TR-69, OR HEPS9100
14 LED's (RED)
14 RESISTORS (R_1)
14 RESISTORS (R_2)
1 RESISTOR (R_3) COMMON TO ALL CIRCUITS
1 14-PIN IC TEST CLIP (PROTO CLIP PC14)
1 14-WIRE RIBBON CABLE

Table 3-4:   Parts list for a 14-pin logic monitor built using the Darlington transistor circuit shown in Figure 3-20

A 16-Pin IC Monitor Using IC's

The basic circuit and a parts list for a logic monitor using a 7406 hex inverter are shown in Figure 3-21.

Figure 3-22 shows a 14-pin test clip and 14-wire ribbon cable with leads cut and bent for easy connection to a breadboard.

How to Test Your Logic Monitor

Step 1: Select a known-to-be-good **IC** (one in a properly operating circuit). Be sure you have the **IC**'s recommended operating conditions, electrical characteristics, etc.; i.e., the spec sheets (see Table 3-1 for an example). The **IC** should have the same number of pins as your test clip.

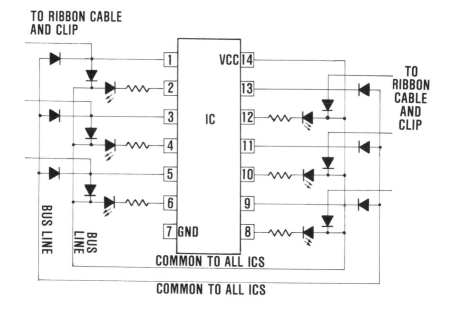

TO RIBBON CABLE AND CLIP

TO RIBBON CABLE AND CLIP

BUS LINE

BUS LINE

COMMON TO ALL ICS

COMMON TO ALL ICS

PARTS LIST

3 - 7406 ICs
32 - Diodes (1N60)
16 - Resistors, 220Ω, ¼ watts
16 - LEDs, red
1 - 16-pin test clip (Proto clip PC-16)
1 - 16-wire ribbon cable

Figure 3-21: Basic circuit and parts list that you can use to build a logic monitor around a 7406 hex inverter IC. Operating power is drawn from the IC under test.

Step 2: Connect the power leads from your logic monitor breadboard to a variable output DC power supply, *set to 0 volts out.* You can use the circuit under test if you do not have a variable power supply.

Step 3: Place the test clip over the **IC** to be tested, making sure the clip is aligned properly. Align the clip mark (usually, a dot) with the index (dot, etc.) on the **IC.**

Step 4: Now slowly raise the variable DC power supply up to the **IC'**s recommended operating voltage (usually about 5 VDC). If

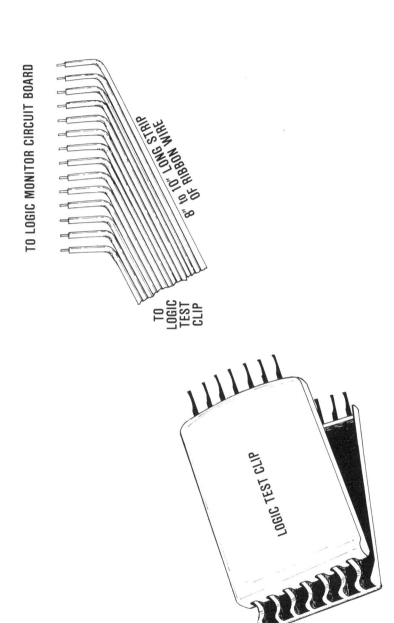

Figure 3-22: Example 14-pin test clip and 14-lead ribbon cable used when constructing a 14-pin digital monitor

there is any question about your breadboard wiring job, monitor the current out of the DC supply. A 7406 inverter is rated for 40 mA of current and the Darlington transistor is rated for 200 mA.

Step 5: Compare the ON and OFF LED's with a truth table or other specs. Do this with several different **IC**'s until you are sure that your monitor is reading properly. *Note:* Most 16-pin **DIP** packages use pin 16 as a DC source pin and pin 8 as ground, but not all. So, be sure you check which pin is V_{cc} and which is ground. You can only check **IC**'s with the same connections for V_{cc} and ground as your clip. To check other **IC**'s, you must change your monitor circuits.

When constructing the electronics, you can use a solderless breadboard, perfboard, or a "home-brew" breadboard of your own design (how to build breadboards is covered in Chapter 4). To see how the logic monitor using a 7406 **IC** works, let's assume pin 1 (Figure 3-21) is connected to the V_{cc} pin of the **IC** under test and pin 2 is connected to a ground pin on the **IC**. The diode at the base of the first LED on the left (LED 1) will conduct. Its bus line will go high. The other diode in the same circuit will not conduct. This keeps its bus line (off pin 1) isolated.

The first inverter inside the 7406 (pins 1 and 2) will see a high at its input, which will produce a low on its output, enabling LED 1 (the V_{cc} indicator) to turn on. The next inverter, pins 3 and 4, will see ground at its input (a low) which, in turn, will produce a high at the output and you will see the light (LED 2) off, indicating ground at the **IC** ground pin. All the rest of the **IC** pins will produce similar results, depending on whether the logic level is high or low. Since most 16-pin **DIP** packages use pin 16 as V_{cc} and pin 8 as ground, this means that, using this hookup, pin 16 will always glow and pin 8 will be dark.

If you decide to use Darlington transistors, as shown in Figure 3-20, you'll need 14, 16, etc., as we have said. The basic circuit works this way. The Darlington transistor drives the LED. Resistor R_2 is used to keep the input at logic level 0, to insure the LED will be cut off. Resistor R_3 is a current-limiting resistor needed by all LED circuits, as has been explained in previous sections. However, if you use this circuit to build your logic monitor, you can use the Resistor R_3 for a current-limiting resistor for all the Darlington transistor circuits you choose to use.

To save yourself a lot of trouble, be sure and mark each LED to show which pin and cable it belongs to. When breadboarding these circuits, it generally is less work if you'll first lay out one circuit (such as the Darlington transistor and its components shown in Figure 3-20) in one corner of the breadboard and work from there. Place your LED's on one side of the perfboard and your wiring on the other side. To save space, mount your resistors vertically. Figure 3-23 shows how this can be done.

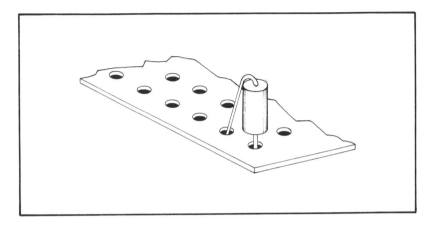

Figure 3-23: How to mount resistors vertically when breadboarding

CHAPTER 4

Practical Building Techniques
for Digital Projects

Every electronics technician/experimenter needs inexpensive "hardware" when building projects. But we are all faced with the same cost problem: if we buy a minumum quantity of electronic hardware, we have to pay a king's ransom before we even start a project; if we purchase in quantity, we end up with our spare parts box full of components we may never use. This chapter offers a solution. In the following pages, you'll find instructions for making a lot of the hardware you can't find or afford.

Also, after your circuit has been thoroughly developed and tested *on paper,* you will need a breadboard. Not just any breadboard—but a breadboard that's inexpensive, easy to build and versatile. Project 4-1, "Building a Breadboard Using Homemade Hardware" will provide you with all these advantages. It will do what the commercial unit does—and a great deal more than most. This "homebrew" breadboard can be used for all types of **IC's: TTL, DTL, RTL, MOS, CMOS,** etc., as well as for discrete transistors.

Included in the following building techniques for digital projects are practical step-by-step procedures for converting schematics to PC patterns, tips for quick prototyping and many other money-saving ideas.

106

Preplanning of a Logic Project

After your circuit has been thoroughly developed and tested on the breadboard you build in Project 4-1, the next step usually is to construct a prototype printed circuit (PC) board. Preplanning of any electronic project is extremely important. However, when laying out printed circuit boards, not only is it important, it's an *absolute must.* The only way I know of getting around the job is to copy a pre-planned circuit from an electronics magazine, manufacturer's technical bulletin, book, etc. Every electronics technician/experimenter soon finds that laying out a PC board can be either fairly simple or extremely difficult, depending upon the circuit. But try to do the job without preplanning and you'll usually end up with a hopeless mess.

The basic idea is simple. You want a minimum number of crossovers (jumpers) and adequate space for all the components you need to mount on the PC board, yet keep the board as small as possible. Of course, it really isn't that simple. Anyone who has tried to draw an IC socket, transistor, and connector patterns on a scale of 1:1 will quickly tell you, after the first, second, and third try, that there has to be a better way. Fortunately, there is.

There are several so-called "Easy Ways" of making prototype PC boards. For the home experimenter who wants to construct only one or two specifically designed (dedicated) PC boards for a certain application, it's best to check around a bit before buying. Two of the companies that you can write to for catalogs are:

1. Bishop Graphics, Inc. (Circuit-Stik Division), P.O. Box 5007, Westlake Village, Ca 91359
    Ask for information on their *Cut-N-Peel System* and *Circuit Stik* method
2. Rainbow Industries, P.O. Box 3266, Indianapolis, IN, 46206. Ask for information on their *Stamp It-Etch It* system.

Also, etched circuit kits are available at electronic stores such as Radio Shack, etc. I think the Cut-'N'-Peel system produced by Bishop Graphics is best, probably less expensive for the home experimenter (assuming you're building only one PC board), and does not involve the mess of etching required by the Stamp It-Etch It system offered by Rainbow Industries and various electronics stores. However, there are advantages in both systems. For example, you can use the stamps in the Stamp It-Etch It system on paper, which really helps during layout planning. A nice feature of the Circuit-Stik system is that you can have crossovers in the foil pattern.

In winding up this discussion of prototyping PC boards, it should be pointed out that pencil wiring of PC boards (and other circuits) has much going for it, unless you are building a one-in-a-lifetime board (see PC/Wire Wrap Techniques in this chapter). Incidentally, before you commit yourself to *any* project, be sure *all* parts required for your project are available at a price you are willing to pay. Look in the back of electronics magazines, where you'll find numerous companies that will be glad to send you free parts catalogs. Write or call for them during your preplanning.

DC Power Supply Requirements

A regulated DC power supply usually is best for most logic families, with one exception — **CMOS**. The reason that you can usually get away with an unregulated supply with a **CMOS** device is that it will tolerate a wide variation in input and output voltage and, in general, will consume very little power. However, as the frequency of any digital signal increases, there will be an increase in the amount of heat a digital device (**IC**, etc.) has to dissipate. What this means is that you will have to use heat sinks and/or be frequency limited when you use the device.

If you are working a digital circuit card up from scratch, be very careful about power requirements. Remember, one component may require only 100 milliwatts, *but* ten of the same components will require 1 watt! Many **IC**'s (such as 35-watt solid state amplifiers, etc.) must be mounted on heat sinks and will place quite a drain on a DC power supply. Therefore, total dissipation per card must be carefully considered regardless of the application. A few points to remember when working with DC power supplies are:

1. In actual use, the effective regulation of a supply may be considerably degraded by the resistance of the connecting leads between the supply and its load. *Short heavy-guage leads are best.*

2. The current rating of a regulated supply is only valid when the operating temperature of the supply is maintained at the level suggested by the manufacturer. *Don't overload a regulated DC power supply.*

3. When you are breadboarding or designing a PC board, high current devices should be located as close as possible to the V + and ground connections.

4. When you have more than one PC board assembly in your project, heavy bus structure should be used to connect the V + and ground lines to all boards. In some cases, you may even have to use separate ground leads to high power circuits.

All About Breadboards and Breadboarding

As we have said, the first step in building a logic system is to lay it out on paper. The next step is to *breadboard* or *prototype* your layout. A modern breadboard is any device (purchased or homemade, such as Project 4-1) that allows electronic components, for experimental work, to be fastened temporarily to a board. Usually, you'll find the commercial models have spring clips, embedded beneath the surface of a perforated block, for the purpose of making the temporary connections. The circuit elements that you want to be electrically connected are inserted by applying pressure on the component leads (for example, an **IC**, etc.). On the other hand, a prototype board usually has all components more or less fixed on the board using soldered or wire-wrap techniques. It is true that components can be removed and exchanged on prototype boards although, generally, not as easily as with the breadboards.

If you are working with a fairly complex circuit, 1) lay it out on paper, 2) breadboard your circuit, 3) then prototype the system on a PC board, using solder or wire wrap. First, we'll look at the characteristics and do's and don'ts of a breadboard system, then, a little later, we'll do the same for prototyping systems.

As an example of a commercial breadboard system, let's use the products of the Vector Electronic Co., Inc., 12460 Gladstone Ave., Sylmar, CA, 91342. Table 4-1 shows this company's breadboarding system characteristics.

HOLE PATTERN	SIZE	INDEXED	BUS STRIPS	MAXIMUM WIRE SIZE	COMPONENT NAMES
0.1″ × 0.1″	8 × 4 to 24 × 4 FOR MOUNTING ON 0.1″ × 0.1″ BOARD)	YES	SEPARATE 1 × 4 TO 1 × 8	TO NUMBER 20	KLIP-BLOCK KLIP-STRIP KLIP-BUS PATCHBOARD

Table 4-1: Characteristics of the Vector Electronics Co., breadboard system

Referring to Table 4-1, you'll see a heading in the left-hand top of the table labeled *Hole Pattern*. Notice, the hole pattern is given as 0.1″ x 0.1″. This is an important piece of information because it means the company has used a standard hole spacing (not all companies do). *This spacing conforms to the pin spacing of practically all IC's* which means you do not have to use **IC** sockets with any breadboard system using this spacing. Figure 4-1 shows how the system is basically set up. There are separate bus strips and, when used with the Klip-Blok and patch board, etc., it's possible to shift all components to almost any position on the board.

Another heading of particular interest in Table 4-1 is *Wire Size.* This information generally is listed by all manufacturers of breadboarding systems and it represents the maximum gauge wire that you can easily insert into the tie points.

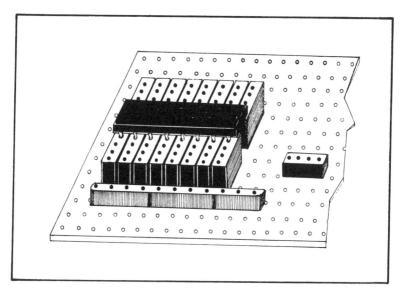

Figure 4-1:   Vector Electronics Co., Klip-Blok, Klip-Strip, Klip-Bus, and patchboard used to make up a breadboard system, and showing a 16-pin IC plugged in

The next heading, *Indexed,* can be a real work-saver, depending on whether the breadboard system is indexed or not (this one is). What is meant by "indexed" is that the breadboard is marked so that you can identify each tie point by referring to some number-letter (or any other identifying marks) combination.

By no means is this example the only breadboard system available. Other than the Vector Electronics, Co., whose address we have already given, you might write to:

1. Saxton Products, Inc., 215 N. Rte. 303, Congers, N.Y., 10920
2. Continental Specialites Corp., Box 1942, New Haven, CT., 06509
3. AP Products, Inc., Box 110-Q, Painesville, OH, 44077

Both design and construction of breadboards have come a long way in the last few years. In the past, constructing circuits for a working model was accomplished by a rat's-nest type setup on the workbench — at least that's what I usually ended up with. Not anymore. Today, anyone working with digital **IC**'s needs a much better setup. A *good* breadboard (home-brew or commercial) should contain seven or eight different circuits. For example:

1. DC power supply for internal circuits.
2. DC power supply for external circuits.
3. Pulse generator (signal source).
4. LED logic state indicator and/or audible logic state indicator.
5. Pulser switch (bounceless).
6. Pulse state monitor.
7. Logic switches (usually 3 or 4).

When using some breadboards, you can still end up with power supplies, switches, indicators and all kinds of gear all over your workbench. There are breadboards (such as Continental Specialties' *Proto Boards*) that are manufactured with built-in power supplies, but we are now talking about a fairly expensive piece of equipment. The home-brew breadboard in Project 4-1 will do what most commercial units will do — plus more.

All of these popular "solderless" breadboards (including the one in Project 4-1) have a distinct problem. There is very high capacitance between the interconnecting strips within the socket. At high frequencies and at very fast switching speeds, capacitive coupling may occur between the holes on the board, or between the closely spaced internal components (wiring, etc.). What this all adds up to is that you should not try to build high-frequency circuits on these solderless breadboards. Restrict your projects to low-frequency circuits in which the switching speeds are low.

There are two other important steps you should follow when using a breadboard. These are:

1. When breadboarding a circuit, try to use the same physical layout you plan to use in the finished project (a PC board, etc.). By doing this, you will reduce the chances of ending up with a job you wish you had never started.

2. The second, and perhaps most important step, is to work out *all* bugs in your circuit while it is set up on the breadboard.

After your circuit has been developed, the next step usually is to hard-wire it. This is not the time for circuit modification—make all changes while the circuit is on the breadboard.

Wiring Techniques

What's the *best way to wire* a project? There are three answers to this question: 1) *traditional wiring;* 2) *wire-wrapping* and, 3) *pen-wiring.* I doubt that traditional wiring methods need much explaining. Almost everyone has cut off a length of wire, stripped both ends, mechanically connected both ends to a solder terminal, and then soldered. This old system is good, and is necessary in some cases. For example, power supplies and buses frequently call for heavier gauge wire than is normally used with low-cost pen-wiring and wire-wrap methods. Both **IC** and transistor circuits are well suited to wire-wrap. A wire-wrap tool that makes your job easier is shown in Figure 4-2.

The wire-wrap tool shown, allows you to slip the tool over the terminals, wrap the wire several turns (seven or more) and then directly run the wire to the next terminal you want to wrap, again slipping the tool over the end of that terminal and wrapping the desired amount of turns around the post. One nice thing about this kind of tool is that when you finish wrapping the last terminal in the series, simply cut the wire and procede to the next circuit path. The actual wrapping is very fast, but speed is rated as medium to high because pins with rectangular cross-sections *must* be used. You *cannot* wrap on round pins or leads because there are no sharp corners to bite into the wire. Also, you *cannot* wire-wrap to capacitors, resistors, etc. for the same reason; i.e., you'll end up with a poor electrical connection because the insulating coating on the wire-wrap wire has not been broken.

FITS OVER
TERMINAL
TO BE
WRAPPED

Figure 4-2: A wiring tool that eliminates pre-cutting and pre-stripping

There are **IC** wire-wrap sockets (usually with three levels of wrapping space), terminals, and pins. Many companies offer these items and, of course, you should check them all to see which are best for you. However, Figure 4-3 shows two examples of terminals and an **IC** mounted on PC board for both solder and wire-wrap methods of connection.

In most cases, you can use the same tool for both wrapping and unwrapping a connection. Therefore, the connection can be changed more easily than soldered joints. When soldering, you can't see how well a solder joint is bonded to a terminal post, but you can inspect a wire-wrap connection simply by looking at it. If it looks all right, it is.

The least expensive of all the wire-wrap tools is the hand-operated type like the one shown in Figure 4-4. This tool strips wire, when used as shown in (A), and will wrap or unwrap wire, if used as shown in (B).

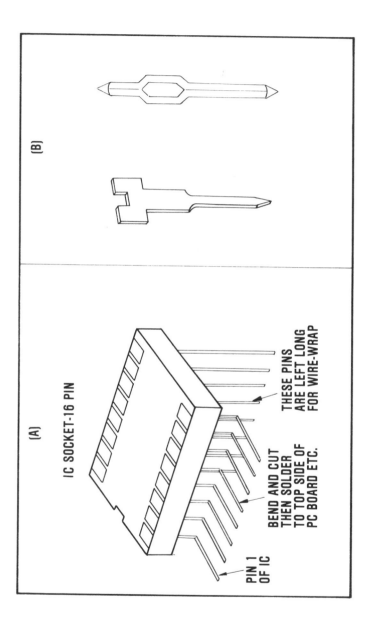

Figure 4-3:   An example of how an IC socket, standard 16-pin, might be mounted on a PC board for both solder and wire-wrap connections. For terminals, you could use either of the pins shown in (B)

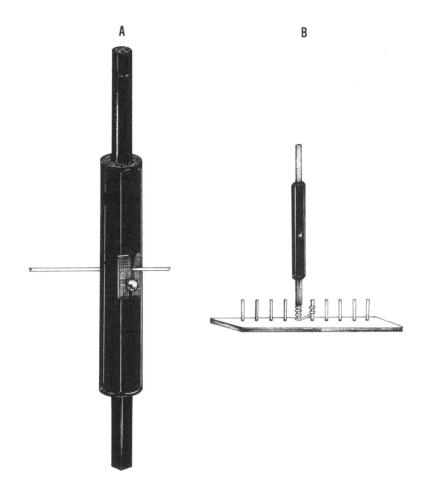

Figure 4-4: Wire-wrap tool that is inexpensive and simple. Each wire must
be cut to length and stripped by hand

Pen-wiring, the third process we listed in the beginning of this section, is especially useful for building projects and may be the one you'll want to use. It is quick and easy and, best of all, *no special* port or terminals (such as the one shown in Figure 4-3) are needed, although you might find they are a help as feed-throughs, etc.

Basically, you "hook up" the wire the same way as has been explained, i.e., wrap a few turns (two or three) around a post (can be resistor leads, capacitor leads, round or square post, or whatever), move on to the next terminal, make two or three turns around it, and then procede on to the next point. However, using this system, you *do not remove* the hook-up wire's insulation. The insulation on the wire used with a pen-wiring tool is made of a special composition that will vaporize when you touch it with a hot soldering iron (minimum temperature is 400°F).

You'll find that pen-wiring is hard to beat. After you finish wiring your project, simply touch each joint that you want to solder with your soldering iron, apply solder, and you are ready to try your project. Vero Electronics, Inc. 171 Bridge Rd., Hauppauge, N.Y. 11787, manufactures a kit (model 79-1738K) that contains a pen-wiring tool, wire, etc. plus other PC board tools.

Using and Working with IC Sockets

Looking through this book, you'll find many diagrams of **IC** pinouts (**IC** pin numbers). If the drawing is not labeled to show which view you are looking at, it is always assumed it is a *top view*. In general, this is true throughout the industry; i.e., sales catalogs, technical bulletins, and other technical literature.

Here's what can happen if you don't pay particular attention to which view is presented. You very carefully install some sockets for **IC**'s, then wire your project for the *mirror image* of the **IC** pinout! How does this come about? Very easily. All you have to do is make the mistake of laying out the bottom of a PC board, etc., using a top view pinout diagram. Or you can also end up in the same kind of mess if a negative is inadvertently flipped before the etching process is begun. The only way I know to try to correct a fouled-up job like this is to unsolder the socket and solder the **IC** to the bottom of the PC board. Or use an old trick from electron tube techniques . . . make a tube socket adapter (in this case, an **IC** socket adapter). To do this, mount a corrected pinout socket above (piggybacked) the erroneous pinout socket, both interconnected (incorrect pins to corrected pins) with short lengths of hook-up wire. Incidentally, it will make your job easier when wiring the two sockets together if you will use one color of hook-up wire for one-half (for example, 8 pins on a 16-pin socket) of the pins and another color for the other half. Also, the socket/socket-adapter combination can be made more professional looking if you cement small plastic strips on the sides of the finished product.

To install an **IC** simply follow these steps:

1. *Disconnect the circuit power!*
2. Check to be positive the **IC** is facing the right direction before trying to insert it. In most cases, the **IC**

will have some form of identifying mark to show you which pin is number 1.

3. Next, start with one side of the **IC** pins. Line this side of pins up with the socket holes. Press slightly until the pins just start to enter the holes in the socket.

4. After you have all the pins slightly started, use a rocking motion and press the **IC** on into the socket until it is firmly in place.

It's a little bit more difficult to remove an **IC** than it is to insert it. As with electronic tubes, sometimes the unit is hot, and sometimes it is just simply hard to get out. One trick is to make some form of puller. For example, some technicians make a puller by taking a pair

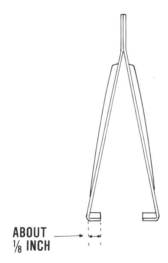

ABOUT
⅛ INCH

Figure 4-5: Home-brew IC puller

of long tweezers and bending the ends inward about ⅛ inch. See Figure 4-5. Of course there are several commercially made **IC** pullers available at most electronic parts stores, but if you don't have one, the tweezers will get the job done.

Another on-the-spot puller is a flat-bladed screwdriver, which is probably the most widely used method of all. It is quick, handy, and inexpensive. Using the screwdriver method requires only that you insert the tip of the screwdriver blade under one end of the **IC** and simply pry up. Go to the other end of the **IC** and do the same. Keep doing this, back and forth, until you have the **IC** out of the socket. *Note: Always disconnect power before removing or inserting any IC!*

How to Plug an IC Test Clip
Directly into an IC Socket

Figure 4-6 shows a rig that you can make up and use to connect an **IC** test clip directly to an **IC** socket. It is inexpensive and can be very handy during troubleshooting. For example, to check for a missing input signal, simply turn off power, pull the suspect **IC**, plug in the adapter/monitor and re-apply power. The test clip will show you whether you do, or do not, have power, input signal, etc.

Converting Schematics to PC Patterns

We have pointed out the fact that, by using materials supplied by companies such as Bishop Graphics, you can make dedicated PC boards with very little effort. Their system consists of epoxy-glass boards, plain or drilled, plus pressure sensitive PC board copper pat-

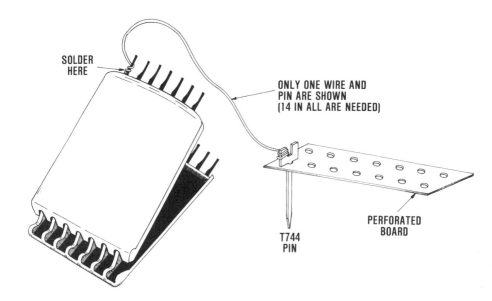

Figure 4-6: IC Test clip (14, 16, or more pins) modified to connect directly to an IC socket

terns, and all types of sockets. However, after you have completed your circuit design on paper, breadboarded the circuit and, very important, *thoroughly checked out its operation,* then, and only then, can you draw a *completed* schematic. At this point, you are ready to start your PC board design by redrawing the schematic diagram as it will be laid out physically on the PC board. To illustrate the layout procedure, we will use the probe circuit diagram shown in Figure 3-17, for Project 3-1. Figure 4-7 is the schematic for this particular probe.

After you have your working sketch, your next step is to start your layout by drawing the **IC**'s as they physically appear, with their pins pointing toward you. Then draw in the other components near the **IC**'s to which they will be connected. However, because we are using a digital probe handle, your drawing should be as compact as possible . . . preferably just slightly wider than the **IC**'s and, of course, not longer than the probe case you plan to use.

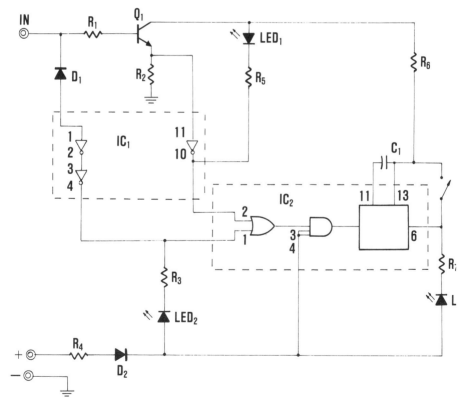

Figure 4-7:   Example of circuit for which a PC board is to be made. See Project 3-1 for circuit details

If your are going to use pre-etched pressure sensitive copper patterns to make your board, be sure all components are included and placed in the best position for easy copper tape interconnections. As you draw in each component (see Figure 4-8) and complete each interconnection, be sure to keep a record of your progress on the schematic shown in Figure 4-7 (or, better yet, on tracing paper or a photocopy). Also, for each connection, make a dot. For example, pins 3 and 4 of **IC** 2 and the connections of LED's 2 and 3 should each have a dot. Figure 4-8 is a suggested layout for the probe components.

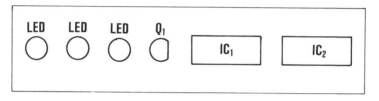

Figure 4-8: Suggested component layout for the PC board discussed in the text

The next step is to redraw the wiring diagram as it will be laid out physically and include the resistors, diodes, etc. Figure 4-9 is an example of how this might be done. Notice, in this case, we have used both sides of the mounting board. You can use pre-etched pressure-sensitive component mounting, an etching process, or hard wiring. But, as we have said, pre-etched (or you do the etching) your board will be smaller than if you use hard wiring.

If connecting lines cannot be made without crossing previously drawn lines, try to rearrange the components and/or re-route your lines to eliminate the crossovers, if you are using the etching process. However, if you try several times to eliminate crossovers and fail, you must use jumpers. Perhaps the easiest way to make crossovers is by using copper tape and then insulating mylar tape between the two copper tapes. Figure 4-10 is an example of how copper tape is used during PC board construction.

As you work up your layout, plot your progress on either the rough drawing (Figure 4-9) or a piece of tracing paper taped over your rough layout. For spacing between component lead pad centers (see Figure 4-10), copper tape conductors, etc., take the dimensions directly from the actual materials and components you will be using in your project. Your final drawing should be laid out *exactly* to

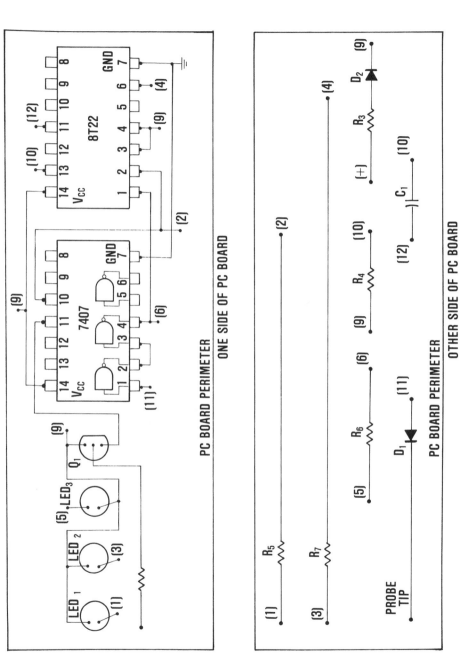

Figure 4-9: Figure 4-7 redrawn as it will be laid out physically

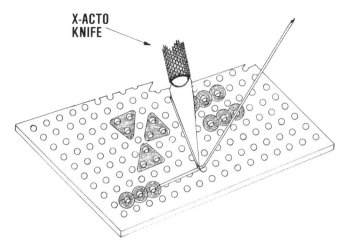

Figure 4-10: Making a dedicated PC board without etching. All patterns shown are pressure-sensitive, as is the copper tape being applied

scale, especially if you are making an etching and drilling guide. You will find that using 10 x 10 squares-per-inch graph paper for accurate placement of pads, etc., will make your job much easier.

Tips for Quick Prototyping

Figure 4-10 shows some examples of press-on copper patterns that you can use to make up an electronic circuit on a board. Use of these patterns will make your prototyping job quick, easy, and reliable. If you are like me, you want to be able to correct mistakes. These patterns, traces, and so on, can be lifted and repositioned (if you do it carefully) within a short time after putting them down. In fact, you also can lift the circuit patterns, tape, etc., even after the adhesive has set. If you do this, the lifted patterns cannot be re-used.

There are many types of patterns you can use. For example, there are pressure sensitive copper tape, donut pads, elbows, and Cut-N-Peel sheets. Some typical applications suggested by Bishop Graphics are:

Pressure-Sensitive Terminals (donut pad)

1. Use when you need to mount a discrete component. They will help you make quick and accurate locations for drilling hole centers.

2. Save PC board construction time with donut pads when mounting axial and radial lead components such as capacitors, resistors, and diodes.

3. These pads are ideal for repairing or modifying a half-finished, or completely finished, printed circuit board.

Pressure-Sensitive Copper Tape

1. With this tape, you can build a "bridge" right on your board by putting insulating tape across existing copper traces and then put the new copper trace right on top of the old trace. You eliminate the job of soldering a jumper wire later (and you *don't* have to be so careful trying to get a perfect design the first try).

2. You will find that this tape is ideal for quick repairing of broken conductor traces on a PC board.

3. Want to modify your PC board? No problem. Simply put down the new adhesive parts, use copper tape for the interconnections and you'll probably be able to salvage an otherwise useless board.

Today, there are pressure-sensitive tapes, insulation sheets, copper sheets, drilled boards, and entire prototyping kits that reduce the whole process of making a one-of-a-kind PC board to nothing but a simple workbench project.

Money-Saving Suggestions for Building Projects

Custom-made electronic hardware is expensive, right? Not necessarily. Why not make the hardware you can't find or can't afford? For example, there are several *push-in* terminals on the market that you can modify to fit your needs. Figure 4-11 shows a couple that can be changed without too much effort (these illustrations show how the pins would normally be used when inserted into a board).

Figure 4-11 (A) is a push-in terminal for 0.042″ diameter holes and Figure 4-11 (B) is to be mounted in a 1/16″ hole (0.062″ to 0.067″). Modification of either of these terminals is fairly simple. For example, you can either cut away excess material to make a flush mount, or add material (such as metal strips) for additional support or components when space is at a premium. The dashed lines in Figure 4-12 show how you can stack more components using the push-in terminal shown in Figure 4-11 (A). Both of the terminals we

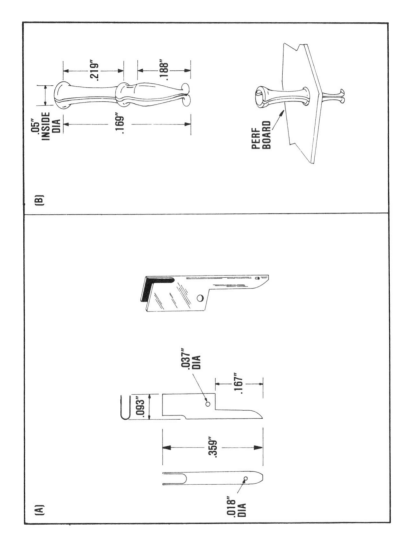

Figure 4-11: Two different type push-in terminals that you can modify to fit your special needs

are using in these example modifications are sold by Keystone Electronics Corp., 49 Bleecker St., N.Y., N.Y., and several other electronic supply companies.

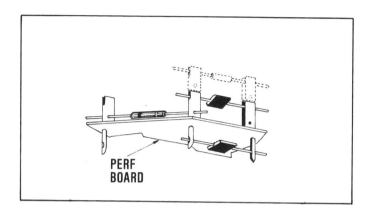

Figure 4-12: How one type of push-in terminal can be made to stack more components, when space is at a premium

Now, let's say you want to make a flush-mount socket on a perforated board (or any other type board). To do this, we can use the miniature tubular terminal shown in Figure 4-11. You modify this terminal by inserting it into your board and then cutting it flush with the board, as shown in Figure 4-13.

After the terminal is securely mounted (several companies sell hand insertion tools at very modest prices), gently tap the terminal until it is all the way flush with the board. Next, clean out any metal burrs so that the desired lead will slide easily into the socket. One last reminder. Don't forget that you can insert these, or any other push-in terminals, into your board upside down. In other words, stack (or flush-mount) the components on the bottom or top side of your board.

Figure 4-14 shows four types of press-fit wire-wrap pins that can be used to make a plug-in terminal strip for discrete components. As an example of how you can do this, let's assume that you want to mount discrete components and then plug in to an **IC** socket. To put it another way, you want to build a *so-called super module.*

First, select a piece of perforated board with the correct hole sizes for the pins you are going to use (in our example, NC "drilled" 0.042-inch diameter type, as shown in Figure 4-14). Cut this board to size (so it will plug into one or more **IC** sockets) and then mount the

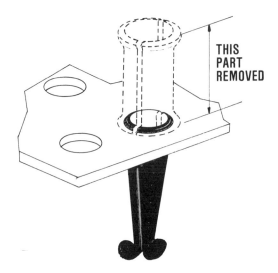

THIS
PART
REMOVED

Figure 4-13: You can make a flush-mounted socket using a stock miniature tubular terminal.

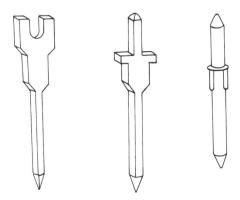

Figure 4-14: Press-fit wire-wrap pins that can be used to make plug-in terminals for mounting strip boards

pins where the **IC** leads would go. After you have the strip cut and pins mounted, it should plug into the **IC** sockets (or socket) you have mounted on your base PC board or breadboard, as shown in Figure 4-15.

You can make and modify all types of specialty hardware at a fraction of what such items usually cost. For instance, need a good probe or screwdriver rack? Next time you are shopping for the family groceries simply pick up an inexpensive multiple-broom holder. They

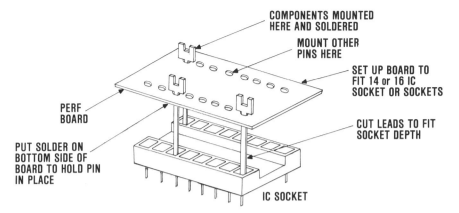

Figure 4-15: "Home-brew" plug-in terminal for discrete components

work very well as a tool rack. Or how about a good low-cost solder wick? If you are having trouble removing solder from IC pins, there is the old trick of using a piece of zip cord (common lamp cord). Simply strip the insulation off the cord and then fray the braided wire, smear some flux on it, and use it for a solder wick. You'll find it will really help when you're trying to pick up hot solder.

PROJECT 4-1:
Building a Breadboard

This project will provide you with a breadboard that is inexpensive, easy to build, and can be used for years to come. The base of your breadboard can be built by using a 6 x 11-inch perforated board and then mounting it on a chassis made of wood. However, construction is greatly simplified by using a plastic parts box of the same dimensions and about 2 inches deep.

In the section titled "All About Breadboards and Breadboarding" at the beginning of this chapter, we listed seven different circuits a good breadboard should have. Therefore, the building plans for each of the circuits will be given here. However, the basic design of the entire breadboard can be easily modified and you can build one, two, or three sections at a time, as you need them. Furthermore, you may want to put additional circuits on your breadboard. You'll find this one is easily tailored to suit your needs. Generally, it's best to build a power supply first. The schematic diagram and parts list for the first (for external circuits) power supply are shown in Figure 4-16.

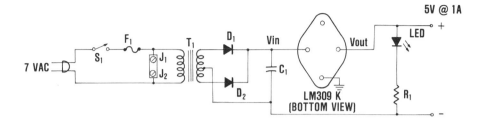

PARTS LIST

1 - AC line cord
1 - SPST on/off toggle switch (S_1)
1 - Fuse holder
1 - 1 Amp fuse (F_1)
1 - 12 VAC CT. Transformer @ 1.2 A (T_1)
1 - LM 390 K voltage regulator
1 - two terminal connector (J_1, J_2)
2 - diodes 1N4002 (D_1, D_2)
1 - 2200 μF @ 35V electrolytic capacitor (C_1)
1 - 270Ω ¼ watt resistor (R_1)
1 - 1.6 - 2.3 LED

Figure 4-16: Schematic diagram and parts list for the external power supply. This supply can be used to supply 1 amp to the circuit you have under development

The power supply is located in the upper left-hand corner of the perforated board or plastic box. The switch (S_1), fuse holder (for F_1), transformer (T_1), and other parts are mounted on the perforated board (or plastic box) with screws, washers, nuts, etc. Mounting holes and even rectangular openings for the switch are easily made in a plastic box — one of the advantages of using plastic. Incidentally, one hole must be enlarged so that you can push the LED through the underside of the box or board. Finally, it's best to apply labels to the switch (ON/OFF), V_{cc} (+ 5V), and ground (−).

A second power supply (for internal circuits) is a light-duty half-wave rectifier. The schematic diagram and parts list for this supply are shown in Figure 4-17. This supply also should be mounted in the upper left-hand corner of the board. The power for it is taken off at the two terminal connector shown in Figure 4-16. *Note: Be sure the connection is placed after the fuse.*

The next most important section of your breadboad is a *pulse generator* (clock). A very good, inexpensive clock can be built using

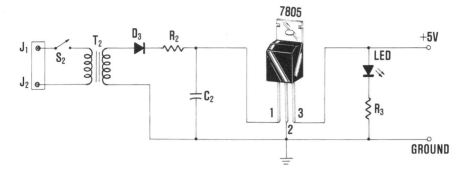

PARTS LIST

1 - SPST on/off toggle switch (S_2)
1 - 12 VAC transformer @ 300 mA (T_2)
1 - Diode, 1N4002 (D_3)
1 - 50Ω, 10W resistor (R_2)
1 - 1000 μF @ 35 V (C_2)
1 - 7805 regulator
1 - 1.6 - 2.3V LED
1 - 270Ω ¼ W resistor (R_3)

Figure 4-17: Power supply schematic diagram and parts list for breadboard internal circuits

the well-known 555 timer **IC**. A 555 clock schematic diagram and parts list are shown in Figure 4-18.

All the components should be mounted so they will be inside a box, except switch S_1, which is mounted through the perforated board. A regular 14-pin **DIP** socket can be used for the 555 **IC**. Placement of the pulse generator must be kept on the outer perimeter of your breadboard (for example, lower left or right-hand corner).

The frequency is changed by selecting one of the various capacitors shown with switch S_1. Of course, you can use additional capacitors or fewer capacitors, should your needs be for other frequencies, or not as many as provided here (0.1, 1, 10 and 100 pulses-per-second). Also, the LED and resistor R_1 may be removed from the circuit and replaced with a 1 k ohm resistor, if you do not want to monitor the state and output of the clock at all times.

Another circuit to include on your breadboard is a *Pulser Switch*. As was explained in Chapter 2, using a manual switch trigger pulse will most likely result in oscillations and result in indeterminate triggering of the circuit under test, if you do not include a debounc-

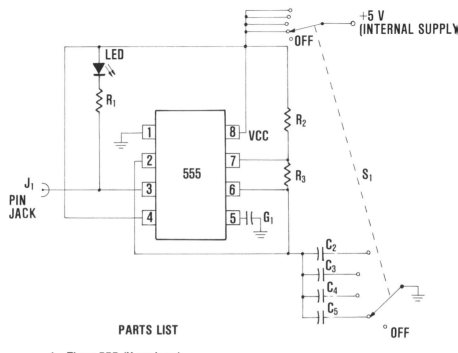

PARTS LIST

1 - Timer 555 (V package)
1 - 270Ω, ¼ watt resistor (R₁)
1 - 1 kΩ, ¼ watt resistor (R₂)
1 - 3.3 meg Ω, ¼ watt resistor (R₃)
1 - 0.1 μF capacitor (C₁)
1 - 1.0 μF capacitor (C₂)
1 - 0.1 μF capacitor (C₃)
1 - 0.01 μF capacitor (C₄)
1 - 0.001 μF capacitor (C₅)
1 - DP 5-position switch (non-shorting) (S₁)
1 - 1.6 - 2.3V LED

Figure 4-18: 555 pulse generator that will produce outputs at about 0.1, 1, 10, and 100 Hz

ing circuit with any type pulser switch. Two gates from an inexpensive 7400 **IC** connected to a SPDT type switch will eliminate the bounce problem. The switch will cause the two gates of the 7400 to change the state of the output. End result – no bounce.

Figure 4-19 shows a pulse switch you can build, that will provide a bounceless switch for your breadboard. Keep this switch and

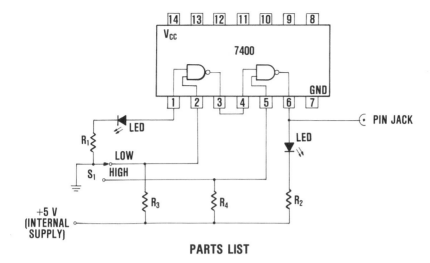

PARTS LIST

1 - Quadruple 2-input positive NAND gate (7400)
2 - 270Ω, ¼ watt resistors (R₁, R₂)
2 - 2.2 kΩ, ¼ watt resistors (R₃, R₄)
1 - LED (green)
  Note: to indicate the state output (lo or hi)
1 - LED (red)
1 - SPDT switch (S₁)
1 - Pin jack (output connection)

Figure 4-19: Schematic diagram and parts list for a pulse switch that will provide bounceless switching for your breadboard

associated circuit on the outer edge of your perforated board. Either top center or bottom center can be used.

If you would like to include a pulse checker on your breadboard, you can use the circuit for the logic memory probe given in Project 3-1, in Chapter 3. Or, simpler yet, if you have already built the probe, or have one, you can use it in place of a built-in checker.

The circuit board is divided into the different circuits that comprise the breadboarding system. All circuits (power supplies, pulse generator, and pulse switch) should be laid out around the outside of the circuit board. Leave the center of the board for one of the numerous rectangular breadboard strips that are sold by several companies. Or you can make a breadboard strip by using tabular solderless pins for jumpers. The ones made by Keystone Electronics Corp., require a 1/16-inch hole and a hand insertion tool to seat the terminal into the circuit board.

During construction, it might be beneficial to lay out the entire breadboard system on top of your circuit board, drill all the holes, deburr the holes, lay on the lettering or legends, spray over the top of the lettering (to protect it), then attach switches, breadboard strips, sockets, etc. You will find that preplanning will really help you get a professional-looking finished product. Figure 4-20 shows a general idea you might like to use for your breadboard system layout.

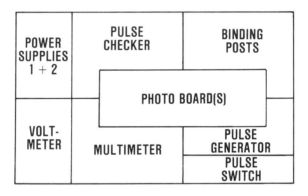

Figure 4-20: Breadboard shown looking at top view. The board on the component side is divided into different circuits of your choosing

Although we have included a voltmeter and milliammeter in our layout, it should be mentioned that almost any type of multimeter can be used to check the presence of the proper DC voltages on the various circuits that you might want to build. However, there are times when built-in meters come in very handy.

CHAPTER 5

How to Use Multivibrators and Flip-Flops In Digital Systems

Multivibrators and flip-flops are two of the most common circuits used in digital electronics. They are very easy to use, low in cost, and available in conventional **TTL** and **CMOS.** This chapter emphasizes their usefulness in practical applications and will provide you with the knowledge that you will need on the job.

In the beginning, the monostable multivibrators' output was simply a one-shot function. Although today's **IC** monostables still provide the one-shot function, their usefulness has been greatly extended. For instance, you can now purchase a 74121 monostable multivibrator, or a 74122 retriggerable monostable multivibrator, or even a 74123 dual retriggerable monostable multivibrator.

Just in case you are not familiar with retriggering, here's how it works. If a second trigger pulse arrives while the output is still high from the first pulse, the output will respond to the latest trigger pulse and remain high. The continuation is for one complete cycle and a stream of input trigger pulses will result in a sustained output pulse that will have a very long duration.

133

EXPERIMENT 5-1
Understanding, Building, and Testing
a Simple Multivibrator

A basic monostable or one-shot multivibrator (a simple pulse generator) can be made by using the popular **CMOS** quad 2-input **NOR** gate, type number CD 4001.

To build this mono, two **NOR** logic gates are direct-coupled by connecting the output of one gate to the input of the other, and the output of the second gate is coupled to the input of the first via a simple R-C time constant network. Figure 5-1 shows the hookup (using one-half of a CD 4001 **IC**) you can use to build the mono.

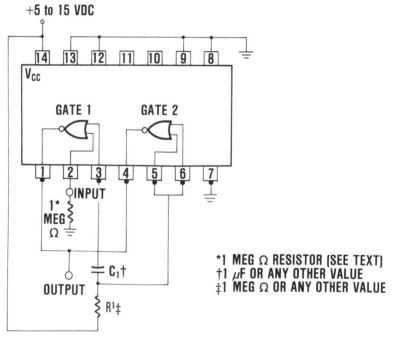

Figure 5-1: Wiring diagram for a basic monostable multivibrator (pulse generator). R_1 can be any value from a few k ohms to several megohms. C_1 can be any value from a few pico F to several hundred micro F.

When you wire the IC as shown in Figure 5-1, you are using the first gate (pins 1, 2, and 3) as a **NOR** logic element, and the second gate (pins 4, 5, and 6) as an inverter, or **NOT** gate. The other two gates are not used. Therefore, pins 8, 9, 12, and 13 should be at ground potential. *Important:* The input of this circuit (pin 2) must always be connected to ground through a 1 megohm resistor, unless a positive trigger pulse is being applied. An equally important point: *Never* apply an input signal without power on. If these rules are not followed, the **IC** may be destroyed.

When you have the **IC** wired as shown in Figure 5-1, and power applied, you should find the input of gate 2 (pins 5 and 6) is at a logic high (through R_1), and the output (pin 4) is at a logic low. Also, you should find a logic low at both input terminals of gate 1 (pins 2 and 3). Pin 1 (the output of gate 1) should be at logic high. Next, since both terminals of capacitor C_1 are at logic high, your multimeter will show that the capacitor is completely discharged.

Now, let's see what happens when you use your logic pulser and apply a positive trigger pulse to the input of gate 1 (pin 2). As soon as you apply the input pulse, you should see the output (pin 4 or 1) go high, remain high for a short time, then quickly return back to its normal low state again. Just how long it will stay high depends on what values you chose for R_1 and C_1, i.e., the R-C time constant.

When you applied the input pulse, it drove the output of the first gate to ground potential, which, in turn, dragged the input of gate 2 with it, via the discharged capacitor C_1. This action caused the output of gate 2 to go high, thus holding the output of gate 1 in a low state even when you removed the trigger pulse. However, in time, there was a change of state because, as soon as gate 1 went low as a result of the trigger pulse, C_1 started to charge through R_1 and the capacitor charging voltage was applied to the input of gate 2 through the R_1, C_1 junction.

After a short time (depending on R_1, C_1 values), the capacitor charge voltage rose to the transfer voltage of gate 2 and, at this point, the output of gate 2 switched quickly back into a logic low state. As the output of gate 2 went to logic low, it caused the output of gate 1 to go to logic high. Capacitor C_1 then quickly discharged through the output of gate 1 and the operating sequence was complete.

A final point to note is that since the output pulse can be made much longer in duration than the input pulse, this type circuit is often called a "pulse stretcher." A mono can be used to stretch a brief pulse so that, among other applications, it can be used to drive a relay.

EXPERIMENT 5-2
Learning About and Constructing a
Digital IC Delay Circuit

Occasions may arise when you need an oscillator with independent control of period and pulse width. The 74122, 74123, **(TTL),** or the 74C221 **(CMOS)** dual-retriggerable monostable multivibrators with *clear,* and using the circuit shown in Figure 5-2, perform this task very well.

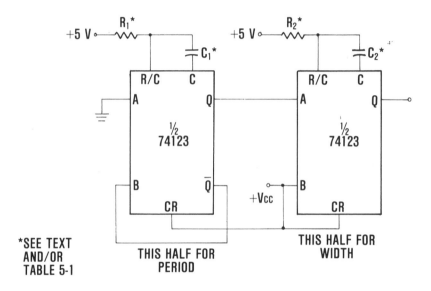

A. H = high level (steady state)
   L = low level (steady state)
   ↑ = transition from low to high level
   ↓ = transition from high to low level
   ⎍ = one high level pulse
   ⎍ = one low level pulse
   ✕ = irrelevant (any input, including transitions)
B. An external timing capacitor may be connected between Cext and Rext/Cxt (positive)

Figure 5-2: This monostable will provide you with complete flexibility in controlling the pulse width, either to lengthen the pulse by retriggering, or to shorten by clearing. Figure 5-3 shows the 74123 IC pin configuration

The IC has an internal timing resistor that allows you to operate the circuit with only an external capacitor, if you wish. However, if you use potentiometers for R_1 and R_2, you can construct a low-cost, wide-range pulse generator with lots of control over the output pulse. The output pulse is primarily a function of the external capacitor and resistor you use. For instance, for a capacitor of greater than 1000 pF, the output pulse width (t_w) is defined as:

$$t_w = 0.32 \, R_T \, C_{ext} \, (1 + 0.7/RT)$$

where:

R_T is in k ohms (either internal or external timing resistor)
C_{ext} is in pF
t_w is in ns

For pulse widths when C_{ext} is less than 1000 pF, see Table 5-1.

OUTPUT PULSE WIDTH VS EXTERNAL TIMING CAPACITANCE

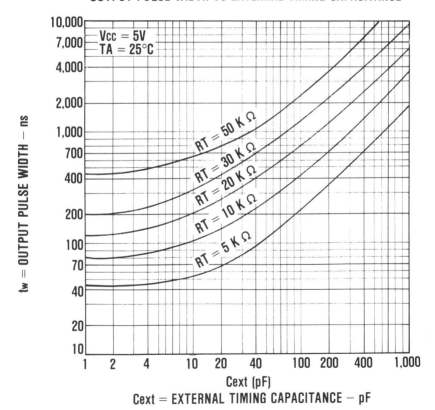

Table 5-1:   Output pulse width Vs external timing capacitance (courtesy Signetics)

This **IC** is also retriggerable, as we have said, and like other retriggerable monostable vibrators, it will respond to inputs that arrive while the output is still high from the preceding trigger. This means you can have a train of inputs that will hold the output high until you stop the train of input pulses. When you wire up the circuit shown in Figure 5-2, you will need the 74123 pin configurations. Figure 5-3 shows the **IC** package with all pins identified.

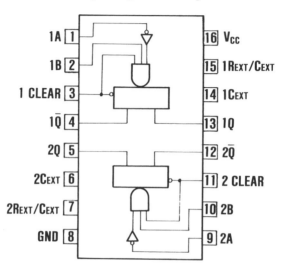

Figure 5-3:   Pin configuration for the 74123 IC

INPUTS		OUTPUTS	
A	B	Q	Q̄
H	X	L	X
X	L	L	H
L	↑	⊓	⊔
↓	H	⊓	⊔

A. H = high level (steady state)
   L = low level (steady state)
   ↑ = transition from low to high level
   ↓ = transition from high to low level
   ⊓= one high level pulse
   ⊔= one low level pulse
   × = irrelevant (any input, including transitions)
B. An external timing capacitor may be connected between Cext and Rext/Cxt (positive)

Table 5-2:   Truth table for a 74123 IC used in experiment 5-2

After you have the **IC** mounted on your breadbc wired for operation, your next step is to run tests on the cir do this, you will need a truth table. Inputs and outputs for th are shown in Table 5-2.

Two other manufacturers' tables that you should have to run this experiment are the electrical characteristics and switching characteristics. Tables 5-3 and 5-4 include this information.

The greatest single source of problems in this experiment is false triggering and the next most frequent problem is no triggering at all. **IC** monostables are very fast and if they can find a way to trigger themselves, believe me, they will. Therefore, when you are breadboarding this circuit, keep your input lines as short as possible. Above all, to prevent stray coupling, keep the input lines as far away as practical from all other lines.

If you have troubles, use a scope with its ground lead connected to your circuit power supply ground. Take a look at your signal ground line to make sure it really is ground. There should not be any detectable voltage or noise (you'll see grass across the scope face, rather than a pure straight zero signal line).

EXPERIMENT 5-3
Constructing and Testing a
555 IC Astable Multivibrator

Although the 555 is a timer **IC**, it can be used as a one-shot, free-running (astable), or gated multivibrator. In fact, this **IC** can be used in a variety of applications; for example, a model traffic light, or to control a model train, car, etc. Figure 5-4 is a 555 timer function diagram that you can use as a guide when working with this **IC**.

For those who like mathematics, there are four defining equations for the free-running mode when using this **IC**. These are:

$$\text{duty cycle} = t_2/t_1 = (R_1 + R_2) / (R_1 + 2R_2)$$
$$\text{output high time } (t_1) = 0.693 (R_1 + R_2)C$$
$$\text{output low time } (t_2) = 0.693 R_2C$$
$$\text{total time } (t_1 + t_2) = 0.693 (R_1 + 2R_2)C$$

If you wish to try a 555 in the free-running mode, Figure 5-5 shows the basic hookup. The frequency of the circuit can be calculated by using the formula,

$$t = 1/T = 1.44/(R_1 + 2R_2) C$$

ELECTRICAL CHARACTERISTICS

PARAMETER		TEST CONDITIONS*	MIN	TYP**	MAX	UNIT
V_{IH}	High-level input voltage		2			V
V_{IL}	Low-level input voltage				0.8	V
V_I	Input clamp voltage	$I_I = -12mA$, $V_{CC} = MIN$			-1.5	V
V_{OH}	High-level output voltage	$I_{OH} = -800\mu A$, $V_{CC} = MIN$, See note 1	2.4			V
V_{OL}	Low-level output voltage	$I_{OL} = 16mA$, $V_{CC} = MIN$, See note 1		0.22	0.4	V
I_I	Input current at maximum input voltage	$V_I = 5.5$ V, $V_{CC} = MAX$			1	mA
I_{IH}	High-level input current data inputs / clear input	$V_I = 2.4$ V, $V_{CC} = MAX$			40 / 80	μA
I_{IL}	Low-level input current data inputs / clear input	$V_I = 0.4$ V, $V_{CC} = MAX$			-1.6 / -3.2	mA
I_{OS}	Short-circuit output current†	See note 1, $V_{CC} = MAX$	-10		-40	mA
I_{CC}	Supply current (quiescent or triggered)	$V_{CC} = MAX$, See notes 2,3		46	66	mA

1. Ground C_{ext} to measure V_{OH} at Q, V_{OL} at Q, or I_{OS} at Q. C_{ext} is open to measure V_{OH} at Q, V_{OL} at Q, or I_{OS} at Q.
2. Quiescent I_{CC} is measured [after clearing] with 2.4V applied to all clear and A inputs, B inputs grounded, all outputs open.
3. I_{CC} is measured in the triggered state with 2.4V applied to all clear and B inputs. A inputs grounded, all outputs open. $C_{ext} = 0.02\mu F$, and $R_{ext} = 25$ kΩ. R_{int} is open.
* For conditions shown as MIN or MAX, use the value specified under recommended operating conditions.
** All typical values are at $V_{CC} = 5V$, $T_A = 25°C$.
† Not more than one output should be stored at a time.

Table 5-3: Electrical characteristics for the 74123 IC

SWITCHING CHARACTERISTICS

	PARAMETER	TEST CONDITIONS	MIN	TYP	MAX	UNIT
t_{PLH}	Propagation delay time, low-to-high-level Q output, from either A input	$C_{ext} = 0$ $C_L = 15pF$ $R_{ext} = 5K$ $R_L = 400$		22	33	ns
t_{PLH}	Propagation delay time, low-to-high-level Q output, from either B input			19	28	ns
t_{PHL}	Propagation delay time, high-to-low-level $\overline{Q}$ output, from either A input			30	40	ns
t_{PHL}	Propagation delay time, high-to-low-level $\overline{Q}$ output, from either B input			27	36	ns
t_{PHL}	Propagation delay time, high-to-low-level Q output, from clear input			18	27	ns
t_{PLH}	Propagation delay time, low-to-high-level $\overline{Q}$ output, from clear input			30	40	ns
$t_w(min)$	Minimum width of Q output pulse	$C_{ext} = 1000pF$ $C_L = 15pF$ $R_{ext} = 10k$ $R_L = 400$		45	65	ns
t_w	Width of Q output pulse		3.08	3.42	3.76	μs

Table 5-4: Switching characteristics for the 74123 IC

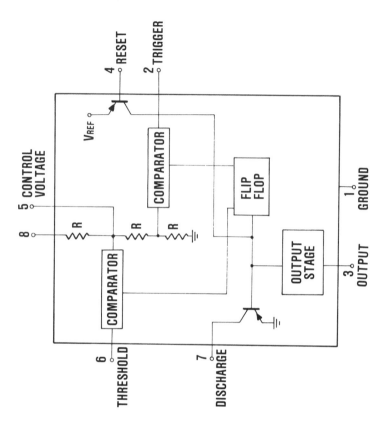

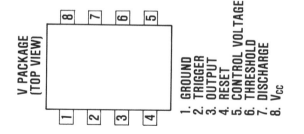

Figure 5-4: 555 timer block diagram and V package, top view

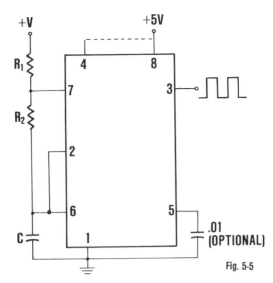

Fig. 5-5

Figure 5-5:   Basic connections for using a 555 timer IC as an astable multivibrator. See text for R_1, R_2, and C values

As you can see by referring to the formula for frequency, if trimmer potentiometers are used for both R_1 and R_2, the frequency (and duty cycle) can be trimmed to your exact requirements. By connecting the output of the basic multivibrator to two LEDs, as shown in Figure 5-6, you can monitor the output or use the circuit as a model traffic light, etc.

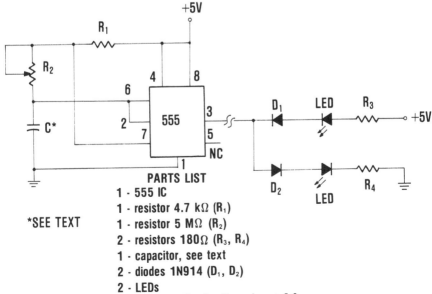

PARTS LIST
1 - 555 IC
1 - resistor 4.7 kΩ (R_1)
1 - resistor 5 MΩ (R_2)
2 - resistors 180Ω (R_3, R_4)
1 - capacitor, see text
2 - diodes 1N914 (D_1, D_2)
2 - LEDs

Figure 5-6:   Schematic and parts list for Experiment 5-3

With any multivibrator, as you know, you'll have an on and off output waveform. However, if it is operating at a frequency, over 20 Hz, your persistence of vision will make it appear that the two LEDs shown in Figure 5-6 are continuously on. Therefore, in order to see the switching action, you will have to set the operation of the circuit at a very low frequency. This (increasing or decreasing frequency) is merely a matter of changing the value of the capacitor labeled C. You can start off with a capacitor value of 0.022 μF, which should permit you to see a definite on-off action.

EXPERIMENT 5-4
Monostable Operation of the 555 IC

In this mode of operation, you will learn how the 555 acts when it is wired as a one-shot multivibrator. The wiring of the IC is simple, as shown in Figure 5-7. The built-in discharge stage at pin 7 of the 555 has an internal transistor that holds the external capacitor, C, initially at discharge (due to a short circuit action by the transistor).

Now, to start the operation, apply a negative trigger pulse to pin 2 (trigger pin). The flip-flop circuit (see Figure 5-4) is now set, and this releases the short circuit across the external capacitor C. At this instant, you should find the output (pin 3) changed to a high logic level.

The capacitor now starts to charge. When this charge voltage reaches about ⅔ the voltage you are using for V_{cc} (5 to 15 V), the 555 comparator (see Figure 5-4) resets the flip-flop, which will then discharge the capacitor, C, very quickly and drive the output to a low state. You should be able to view this with a voltmeter, logic probe, or scope. If you are checking with a scope, Figure 5-8 shows the approximate wave patterns you should see at the input, output, and the capacitor voltage. The exact waveforms will depend on what values you use for R and C, the scope sweep time, etc.

Once you trigger this circuit, it will remain in this state until the time you have it set to has elapsed, even if you trigger it again during the set time interval. You should prove this to yourself during this experiment. You will find that the circuit triggers on a negative going input signal when the level reaches ⅓ of the supply voltage (V_{cc}) you chose to use. However, the time that the output is in the high state depends on what values you chose for R and C. The time that the output is high can be calculated by using the formula $t = 1.1$ RC. As you can see, 1 megohm and 1_μ F will produce a time period of more than a second.

Once you have a time period of about a second or so, you might like to try applying a negative pulse to both pin 4 and pin 2 (trigger) during the pulse period. You must apply the trigger pulse to both pins simultaneously (reset should not be connected to V_{cc} when performing this part of the experiment). This will discharge the external capacitor (C) and start the cycle over again. After completing this

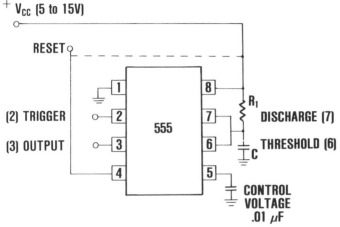

Figure 5-7:   Wiring diagram for using a 555 timer IC in the monostable mode of operation. The capacitor connected to pin 5 is essential to reduce noise. Also, when the reset function is not being used, pin 4 should be tied to pin 8 (V_{cc}) to avoid false triggering

part of the experiment, don't forget to reconnect pin 4 (reset) back to pin 8 (V_{cc}). If you don't, you may have problems with false triggering while trying to operate the circuit as a one-shot.

EXPERIMENT 5-5

Using and Testing a Schmitt-Trigger IC

You will find that a Schmitt-trigger **IC** is one of the most reliable and useful of all possible input schemes. One of the reasons is that its overall effect is to clean up noisy or irregular digital signals so that they can be applied to other logic **IC**'s. Futhermore, Schmitt-trigger gates are widely available at comparatively low cost (about fifty-cents for a 7413).

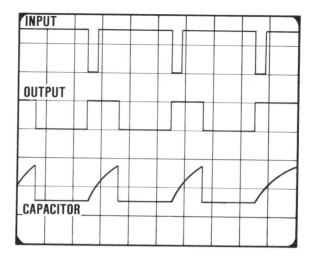

Figure 5-8:   Example waveforms that you can expect to see when using an oscilloscope to view a 555 timer wired for monostable operation

Figure 5-9 shows the pin configuration and logic symbol for the dual 4-input positive-**NAND** Schmitt-triggers 7413. Note that the symbol is that of a logic inverter with a hysteresis loop figure drawn in its center. Normally, there is hysteresis between an upper and lower triggering level. The hysteresis, or backlash, which is the difference between the two threshold levels, typically is 800 mV for this **IC.**

As an example of what a Schmitt-trigger can do for you, let's assume that a digital pulse has been sent through a noisy circuit such as a long cable, and arrives at the input looking like the one shown in Figure 5-10. Next, look at the logic levels emerging from the **IC** — the signal labeled *output.* Notice, it is perfectly clean and ready for application to the next stage. *Note:* A Schmitt-trigger gate cannot be triggered from straight DC levels.

There are a number of methods you can use for breadboarding this Schmitt-trigger (i.e., the 7413). As has been explained in previous sections, there are a variety of **IC** sockets on the market. The type you use will determine how you mount and/or wire the **IC** for this experiment. The most important information you need after the circuit is mounted is the electrical characteristics and the operating conditions for the 7413 **IC.** The recommended operating conditions are shown in Table 5-5. *Note: Leave the power off until the breadboarding is complete.*

PARAMETER		MIN	NOM	MAX	UNIT
SUPPLY VOLTAGE		4.75	5	5.25	V
FAN-OUT FROM EACH OUTPUT	HIGH LOGIC LEVEL			20	
	LOW LOGIC LEVEL			10	
OPERATING FREE—AIR TEMPERATURE, T_A		0	25	70	°C
MAXIMUM INPUT RISE AND FALL TIMES			NO RESTRICTION		

Table 5-5:   Recommended operating conditions for a dual NAND Schmitt-trigger (7413 IC)

To test the device, we will use a simple scheme that will change common house current (60 Hz) into a digital signal (1 or 0 binary). Figure 5-11 shows a schematic diagram and parts list for a test setup that will perform this task.

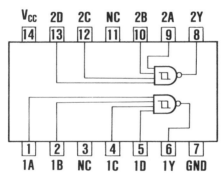

Figure 5-9:   Pin configuration and logic symbol for the 7413 dual 4-input positive NAND Schmitt-triggers. These gates were specifically designed for cleaning up noise or irregular digital signals

In order to work with this **IC**, you need to have at least a nodding acquaintance with the input circuits. Logically, each circuit functions as a 4-input **NAND** gate but, because of the Schmitt action, the gate has different input threshold levels for positive and negative-going signals. Figure 5-12 shows a schematic for a single gate input.

The special Schmitt-trigger levels are designated $V_T +$ and $V_T -$. The V_T voltage level is the input level at which a positive-going signal causes the trigger circuit output to switch from logic 1 to logic 0. The $V_T -$ level is the input level at which the negative-going signal

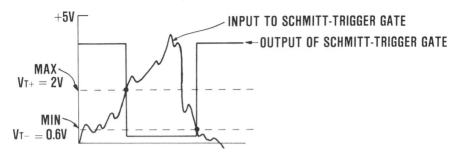

Figure 5-10: Irregular signal input to a Schmitt-trigger gate and its cleaned-up output signal

causes the output to switch from logic 0 to logic 1 and to be ready for the next cycle. There are minimum and maximum voltages for both $V_r +$ and $V_r -$. These are given for the 7413 **IC** , in Table 5-6.

What happens between the two levels (shown as $V_{T+} - V_{T-}$, in Table 5-6) in the *hysteresis* interval, depends on the state of the output as the input signal enters the interval. If the output is at logic 1 when your injected input signal enters the hysteresis level, it will stay at logic 1 until your input signal exceeds the V_{T+} threshold level. On the other hand, if the output happens to be at logic level 0 during the time your input signal enters the hysteresis interval, you'll find that the output will remain at logic 0 until the signal drops to about 0.8 volts (assuming $V_{CC} = 5V$). Those threshold actions are why the Schmitt-trigger is able to clean up noise or irregular signals and why the **IC** can be used to change the 60 Hz line current to the logic levels produced by the circuit shown in Figure 5-11.

PARAMETER		TEST CONDITIONS	MIN	TYPICAL	MAX	UNIT
V_{T+}	POSITIVE-GOING THRESHOLD VOLTAGE	$\overline{V_{CC}} = 5V$	1.5	1.7	2	V
V_{T-}	NEGATIVE-GOING THRESHOLD VOLTAGE	$V_{CC} = 5V$	0.6	0.9	1.1	
$V_{T+} - V_{T-}$	HYSTERESIS	$V_{CC} = 5V$	0.4	0.8		V

Table 5-6: Voltage values of the 7413 IC thresholds designated $V_r +$ and $V_r -$.

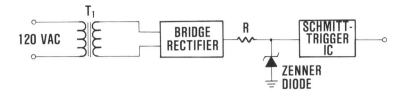

1 transformer, step down, 120 VAC to 6.3 VAC (T_1)
1 bridge rectifier, almost any full-wave bridge rectifier assembly (for example, a 3N246)
1 resistor, 220Ω (R)
1 Schmitt-trigger, 7413
1 zener diode, 5.1 V, ½ watt

Figure 5-11: Test setup and parts list for breadboarding a test circuit for the 7413 Schmitt-trigger IC

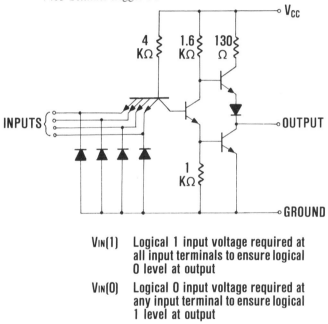

$V_{IN}(1)$ Logical 1 input voltage required at all input terminals to ensure logical 0 level at output

$V_{IN}(0)$ Logical 0 input voltage required at any input terminal to ensure logical 1 level at output

Figure 5-12: Schematic of a signal 4-input NAND gate

EXPERIMENT 5-6
How to Use and Test Flip-Flop ICs

If you built the master-slave **J-K** flip-flop shown in Figure 3-9 and ran the troubleshooting experiment described in Chapter 3, you are already familiar with the action of a flip-flop. According to the thinking of many electronic technicians, you can never go wrong using **J-K** flip-flops, and they probably are not too far off the truth. For example, various combinations of these **IC**'s can provide many useful digital counting schemes; ring counters, shift counters, up/down counters, and so on.

For this experiment, we will use the inexpensive 7476 **IC**. You will also need a hex inverter, to construct a trigger pulse bounceless switch. The schematic diagram and parts list are shown in Figure 5-13. After you have the 7476 and 7407 **IC**'s wired up, you can refer to Truth Table 5-7 for an understanding of what happens when you operate the switches.

PRE	CLR	J	K	CLK	Q	Q̄	MODE
0	1	X	X	X	1	0	ASYNCHRONUS PRESET
1	0	X	X	x	0	1	ASYNCHRONUS CLEAR
1	1	1	0	⎍	1	0	SYNCHRONOUS PRESET
1	1	0	1	⎍	0	1	SYNCHRONOUS CLEAR
1	1	0	0	⎍	Qt−1	Q̄t−1	MEMORY
1	1	1	1	⎍	Q̄t−1	Qt−1	TOGGLE

PRE = CLR = 0 IS INVALID
X = DON'T CARE
⎍ = COMPLETE BLOCK WAVEFORM

Table 5-7: Truth table for a 7476 dual J-K master-slave flip-flop with preset and clear

The preset and clear inputs preset the **J-K** flip-flop to a desired state before another operation is begun. These two inputs are referred to as asynchronous inputs because they do not require a transition on the clock input. You should find that the **J** and **K** inputs affect only the Q and Q̄ inputs (LEDs) when you cause a transition to happen (press the clock switch) on the clock input. If the **J** input is 1 and the **K** input is 1, the flip-flop should reset from the previous state when you change the clock input from low to high.

To set the **IC**, apply a 1 to the **J** input and a 0 to the **K** input, then apply a low-to-high transition to the clock input. This operation is synchronous with the clock operation, i.e., with the clock switch.

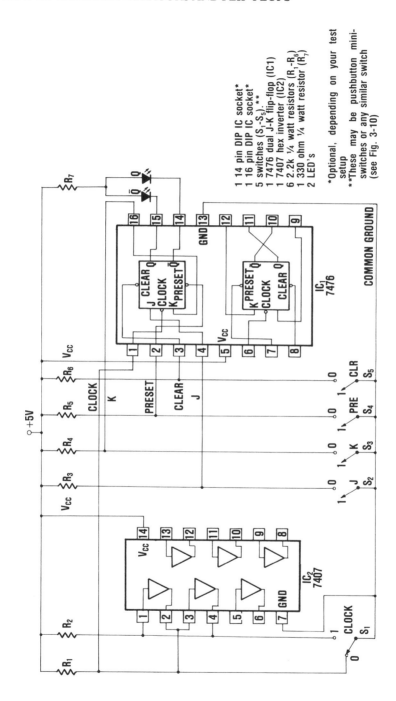

Figure 5-13: Schematic diagram and parts list for Experiment 5-6

1 14 pin DIP IC socket*
1 16 pin DIP IC socket*
5 switches (S_1-S_5),**
1 7476 dual J-K flip-flop (IC1)
1 7407 hex inverter (IC2)
6 2.2k ¼ watt resistors (R_1-R_6)
1 330 ohm ¼ watt resistor (R_7)
2 LED's

*Optional, depending on your test setup
**These may be pushbutton mini-switches or any similar switch (see Fig. 3-10)

The circuit can be put into a memory mode by setting preset = clear = 1 and **J** = **K** = 0. Once you have caused a complete clock pulse to occur, the circuit will remember the states it held at the end of the previous clock pulse. You will also notice (see Table 5-7) that you can operate the flip-flop in a toggle mode (switch back and forth from high to low), as long as you keep preset = clear = **J** = **K** = 1. You'll find that the negative-going edge of each clock pulse will switch the output state.

One last comment . . . setting preset = clear = 0 is an invalid operation and you should not use this mode. In general, you should only change the **J** and **K** inputs while the clock pulse is at logic 0. In fact, if you try to change the **J** and **K** inputs while the clock input switch is in the high position, you will have to cycle the entire operation all over again. You will find that the Q outputs will not show any change until the clock drops to 0 and the circuit is recycled.

Other flip-flops can be tested by using the appropriate truth table and setting up the input switches before the clock pulse is initiated. A truth table for a 7473 (dual **J-K** master-slave), **RS** flip-flop, and for the D-type flip-flop are shown in Chapter 2.

Remember, leave the power off on your test setup (breadboard, etc.) when installing and removing **IC**'s from the sockets. Also, you will find that an oscilloscope is essential, if you want to observe the actual waveforms. However, you can monitor the test signals using the home-brew test equipment found in Chapter 3, Projects 3-1, 3-2, and 3-3.

EXPERIMENT 5-7
How to Determine the Duty Cycle of an
IC Mulitvibrator

In Experiment 5-3 and 5-4, you worked with the 555 **IC** timer. Once again, we will use this **IC**. However, this time we want to determine its duty cycles. Figure 5-14 shows the connections for this experiment.

While you are in the process of building this circuit on your breadboard, it is best to use a standard value of capacitance for C, then calculate the required resistance. You can always use different values of resistance in series, parallel, or combinations, but, as we all know, it is difficult to find odd-ball values of capacitance in a spare parts box. For this experiment, the equation you should use is:

$$\text{duty cycle} = (R_a + R_b) / (R_a + 2R_b) = t_2 / t_1$$

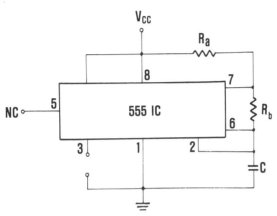

Figure 5-14: Wiring diagram for Experiment 5-7

As you can see from this equation, we are particularly interested in the two resistors labeled R_a and R_b in **Figure 5-14.** Why? Because these two resistors are important if you want to predict or design a certain duty cycle, using the 555 timer in an astable circuit.

Of special interest in the formula for duty cycle is the fact that if you make R_a equal to zero, it places pin 7 at V_{cc} potential and the duty cycle equal to 0.5. *Don't do it!* There is no internal current limiting resistor within the 555 **IC.** Your best bet, in most applications, is to always select a duty cycle of 0.525 or 0.530. To repeat; *the duty cycle of an astable 555 **must** be greater than 50%.*

The circuit shown in Figure 5-14 allows a wide selection of both frequency and duty cycle from a single capacitor. For example, if you use a capacitor value of 0.01 μF and a resistance of 2200 ohms for R_a, you can then vary the value of R_b for several frequencies and duty cycles. For instance, when R_a = 2200 ohms, R_b = 10,000 ohms, the duty cycle = 0.054 or 54%. Setting R_a and R_b both to equal 100,000 ohms, will produce a duty cycle of 200/300, or 66 ⅔%.

If you use trimmer potentiometers for both R_a and R_b, the frequency and duty cycle can be trimmed to your exact requirements. Just remember, *don't set the trimmer R_a to zero resistance.* To be on the safe side, always start with R_a set at high value. *Note:* there are several different formulas for duty cycle (used by the 555 IC manufacturers and other authors) for operation of this **IC** in the astable mode. The equation given in this experiment has been checked and double-checked and found to be correct.

CHAPTER 6

Procedures for Using Digital Counters and Registers

Digital counting circuits are becoming the standard means for measuring almost any variable. Just a few examples are time, frequency, voltage, current, resistance, temperature, and events. But did you know that you can make a decade counter count by some number other than ten? This is just one of the ways this chapter will show you how to get more out of a counter **IC**.

Each of the following experiments is designed to give you a working knowledge of digital counters and registers. This chapter will help you whether you design, repair, or maintain digital **IC** equipment. Or, if you simply would like to expand the possibilities of your workshop, you will find the following information an invaluable aid.

How Digital Counters Count

In general, readouts used and digital equipment normally display in decimal. But, as has been explained in Chapter 2, the counting circuits themselves usually count in binary code. You will remember that the decade **IC** counter automatically resets to 0 after the decimal count of 9, running from 0 to 9, and then repeating. To count to a higher number, the counter output triggers the next largest value. In other words, 09 becomes 10 on the next count, and 199 becomes 200, etc.

155

Now, what if you want the circuit to start back at 0 at some point before it reaches 9? That may sound like a tough question, but it really isn't. In the following pages, you will see how to do this. However, let's first use the popular and inexpensive 7490 decade counter to learn how a basic counter works.

Actually, this is a monolithic decade counter consisting of four dual-rank, master-slave flip-flops internally interconnected to provide a divide-by-two counter and a divide-by-five counter. It should be remembered that while counters such as the 7490 are properly considered binary counting circuits, it is also possible to view them as frequency dividers. This is because, as you'll recall, any toggled flip-flop naturally divides its clock frequency.

Figure 6-1 shows the pin configurations of a 7490 decade counter. Notice that the output from flip-flop A is not internally connected to the succeeding stages. For this reason, the count may be separated into three independent count modes: (1) binary coded decimal decade counter, (2) divide-by-ten, and (3) divide-by-two and divide-by-five. For example, the **IC** can be wired, as shown in Figure 6-2, to form a BCD count sequence.

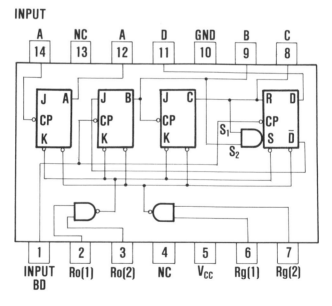

Figure 6-1:   Pin configuration for a 7490 decade counter **IC**

Referring to Figure 6-1, you will notice that pins, 2, 3, 6, and 7 are all inputs to **NAND** gates. These pins are grounded for counting.

Of course, pin 10 must also be grounded because it is the **IC**'s common ground point. *Note:* It is possible to ground only pins 10, 2, or 3, and either 6 or 7 and the circuit will function as a counter. However, for stable operation, it's best to ground them all, as shown.

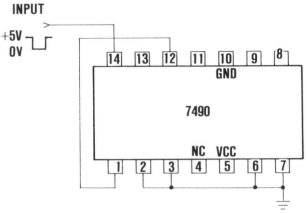

Figure 6-2: How to connect a 7490 IC for a BCD count sequence. *Note:* Output A (Figure 6-1) is connected to input BD for a BCD count.

Next, refer to Table 6-1. Notice that when there is a logical high (about 5 volts) at output A (pin 12), it represents a count of 1.

BCD COUNT SEQUENCE

DECIMAL	OUTPUT			
NUMBER	D	C	B	A
0	0	0	0	0
1	0	0	0	1
2	0	0	1	0
3	0	0	1	1
4	0	1	0	0
5	0	1	0	1
6	0	1	1	0
7	0	1	1	1
8	1	0	0	0
9	1	0	0	1

Table 6-1: Logic truth table for BCD outputs from a 7490 IC, when connected as shown in Figure 6-2.

To put it another way, read under the heading "Decimal Number" the number 1, then read to the right 0001, the binary number 1. Output B (pin 9) is a decimal count of 2, binary number 0010. Output C (pin 8) is a decimal count of 4, binary number 0100, and output D (pin 11) is a decimal count of 8, binary number 1000.

If you wire the 7490 **IC** as shown in Figure 6-2, it should reset to 0 at the next count after 9. Gated direct reset lines are provided (for example, pins 2 and 3 in Figure 6-2) to inhibit count inputs and return all inputs to logic 0. The gated circuitry is designed so that when *both* pins 2 and 3 go high, the **IC** resets to 0. During normal counting, you should find that either pin 2 or 3 (or both) is at a low.

Now, remember that we said you could make a counter count to some number other than 10? To see how this can be done, try this. Connect a jumper wire from pin 9 (B, in Figure 6-1) and remove the ground from both pins 2 and 3 (the gated inputs). Incidentally, when you leave a pin ungrounded, it is the same as if you had connected it to a logic high level. Once you have made these changes, you should find that the **IC** will reset to 0 at the second count; i.e., count from 0 to 1 and back to 0.

If you want the counter to count to 3 (0 to 2), connect pin 2 to pin 9 and pin 3 to pin 12. For a count of 4, connect pin 2 or 3 to pin 8 (don't connect the other gated pin . . . leave it open). For a count of 5, connect pin 8 to one reset pin and the other reset pin to pin 12. For a count of 6, pins 8 and 9 are connected to the reset pins. If you want to count to a higher number, it is sometimes necessary to make the **IC** reset by placing pins C (8), B (9), and A (12) at a high all at the same time. The least expensive way to do this is to use a triple **AND** gate such as the 7411 shown in Figure 2-7, or the 7408 shown in Figure 2-8. If you should have to do this, you must trigger the reset terminals with a logic high when all inputs to the **AND** gate are high. The wiring is: pins 8, 9, and 12 to the **AND** gate inputs (one pin to each of the three **AND** gate inputs); connect the **AND** gate output to the gated pin (pin 2) of the 7490 **IC** and leave pin 3 of the 7490 disconnected. With this setup, the **IC** should count 0 to 6 and reset to 0 on the seventh count.

Can you make the 7490 **IC** do other counts? Sure can. For example, if you want a divide-by-ten count for a frequency synthesizer, or some other application requiring division of a binary count, simply make an external connection (use a jumper wire) between pins 11

(D) and 12 (A). Now, the input count is applied at the BD input (pin 1) and a divide-by-ten square wave is obtained at the output A (pin 12). *Note:* Do not use a jumper wire between pins 2 and 12 (as shown in Figure 6-2) in the divide-by-ten mode of operation. Referring to Figure 6-1, you will notice that there are actually 4 reset inputs, $R_{o[1]}$, $R_{o[2]}$ and $R_{g[1]}$, $R_{g[2]}$. A logic truth table for reset/count is shown in Table 6-2.

RESET/COUNT

RESET INPUTS				OUTPUT			
Ro(1)	Ro(2)	Rg(1)	Rg(2)	D	C	B	A
1	1	0	X	0	0	0	0
1	1	X	0	0	0	0	0
X	X	1	1	1	0	0	1
X	0	X	0	COUNT			
0	X	0	X	COUNT			
0	X	X	0	COUNT			
X	0	0	X	COUNT			

Table 6-2:   Logic truth table for the reset/count inputs of a 7490 decade counter IC

For operation as a divide-by-two, and a divide-by-five counter, you do not have to make any external interconnections. Of course, you must have V_{cc} (min. 4.5 V, nom. 5V, and max. 5.5 V) on pin 5 and ground at pin 10. Flip-flop A is used as a binary element for the divide-by-two function. The BD input (pin 1) is used to obtain binary divide-by-five operation at the B, C, and D outputs. When you are operating the 7490 in this mode, the two counters operate independently, but all four flip-flops are reset simultaneously.

A few paragraphs ago, we said that if you want to count to a higher number, it might be necessary to use a triple input **NAND** gate to feed the 7490. But you can encounter divide-by-sixty counters in clock circuits that use only two 7490 **IC's** and do not need any **NAND** gates or other such devices. Figure 6-3 shows a wiring diagram that you can use to try this experiment.

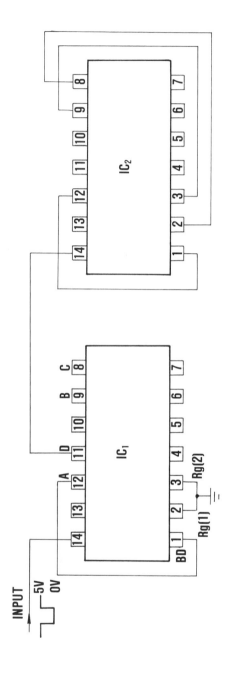

Figure 6-3:   Wiring diagram for a divide-by-sixty counter using two 7490 ICs

IC Binary Counter

When the need for counting events in an electronic system arises, most of us start looking for an **IC** that is both readily available and low cost. The 7493 4-bit binary counter will satisfy these requirements. The **IC** shown in Figure 6-4 consists of a group of four master-slave flip-flops that are internally interconnected to provide a divide-by-two and a divide-by-eight counter. By examining the pin configuration, you will see that the output from flip-flop A is not internally connected to the succeeding flip-flops. Therefore, the counter can be operated in two independent modes . . . as a 4-bit ripple-through counter or a 3-bit ripple-through counter, depending on the external connections you use.

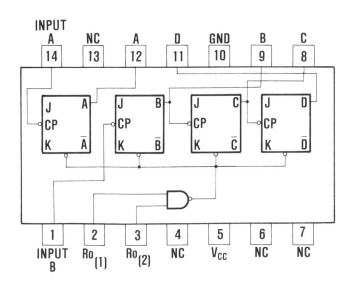

Figure 6-4:   Pin configuration for a 7493 4-bit binary counter

If you want the **IC** to function as a 4-bit ripple-through counter (one pulse ripples along through the flip-flops to provide the outputs), output A must be externally connected to input B. The input count pulses are applied to input A and simultaneous divisions are performed at the outputs, as shown in Table 6-3:

To illustrate how this counter can be used, let's assume that your needs call for a 4-bit up-counter. This **IC** is designed for the job.

COUNT	OUTPUTS			
	D	C	B	A
0	0	0	0	0
1	0	0	0	1
2	0	0	1	0
3	0	0	1	1
4	0	1	0	0
5	0	1	0	1
6	0	1	1	0
7	0	1	1	1
8	1	0	0	0
9	1	0	0	1
10	1	0	1	0
11	1	0	1	1
12	1	1	0	0
13	1	1	0	1
14	1	1	1	0
15	1	1	1	1

Table 6-3:   Truth table for a 7493 IC when used as a 4-bit ripple-through counter

You connect pin 12 to pin 1, pin 4 to V_{CC} (about 5 V), then ground pins 10, 2 (R_{o1}), and 3 (R_{o2}), for a count up to 16. On outputs A, B, C and D, you will have frequency division of 2, 4, 8, and 16. The wiring diagram for this operation is shown in Figure 6-5.

If you want flip-flops B, C, and D to operate as a 3-bit ripple-through counter, you should apply the input pulse to input B. You will find simultaneous frequency divisions of 2, 4 and 8 at outputs B, C, and D. Using this mode of operation permits you to use flip-flop A independently if the reset function occurs at the same time as it does with the reset of the 3-bit ripple-through counter. The gated direct reset line (pins 2 and 3) is provided to stop the count and simultaneously return all four flip-flop outputs back to logical 0.

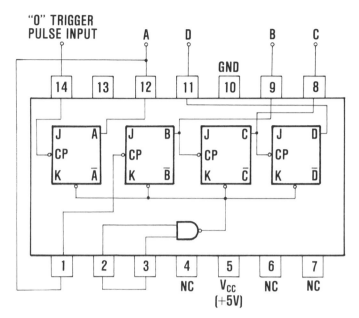

Figure 6-5: Wiring diagram for connecting a 7493 IC for use as a 4-bit binary counter. Supply voltage (V_{cc}) min. 4.75, nom. 5, max. 5.25. The minimum input voltage required to insure logical 1 at any input terminal is 2 V at min. V_{cc}. Max. input voltage for logical 0 at min. V_{cc} is 0.8 V.

EXPERIMENT 6-1
Using and Testing Binary Up/Down Counters

You can use a 74193 4-bit binary up/down counter to count up from 0 to 15 or, by using other inputs, count down from 15 to 0. As shown in Figure 6-6, this up/down counter has an input for count down (pin 4) and another for count up (pin 5).

The outputs of the four master-slave flip-flops contained in the **IC** are triggered by a low-to-high transition of either clock input (pin 4 or 5). Synchronous operation is possible by having all flip-flops clocked simultaneously so that the outputs change state at the same time, when instructed to do so by the steering logic. However, the outputs may also be preset to any state by entering the desired data at the data inputs while the load input is low. In this case, the output will change to agree with what you have entered independent of the count pulses. This feature allows counters to be used as programmable dividers by simply changing the count length with the preset inputs.

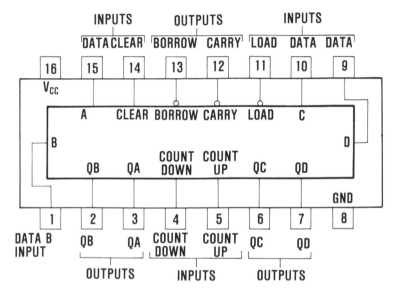

Figure 6-6: Pin configurations for a 74193 synchronous 4-bit binary up/down counter with preset inputs

The two outputs, borrow (pin 13) and carry (pin 12), can be connected to clock-down and clock-up inputs of a subsequent counter that you need to use when working with numbers greater than 15. To put it another way, the 74193 counter **IC** is designed to be cascaded without the need for the external components. Simply connect pins 13 and 14, as explained, and you can cascade both the up and down functions for higher number capabilities. The basic circuit for this experiment is shown in Figure 6-7.

To test the 74193, use the following procedure.

Step 1: Wire in the 5 switches (A, B, C, D, and E).

Step 2: Connect LEDs to output pins number 2, 3, 6, and 7.

Step 3: Insert the **IC** into a socket.

Step 4: Connect + V_{cc} to pin 16, ground to pin 8. Set V_{cc} to 5 volts.

Step 5: Data pins (switches A, B, C and D) are set high for the desired number to be entered (see Figure 6-7).

Step 6: Load pin (switch E) is normally high but you must set it to *low* momentarily, in order to load the **IC** (see Figure 6-7)

Step 7: If you wish to count up, count-down input must be at a high. To count down, count-up input must be high.

Step 8: Pulse either pin 4 (count down) or pin 5 (count up), depending on which way you want to count.

Step 9: Place a logic 1 level pulse on pin 14, to clear the **IC** (see Figure 6-7).

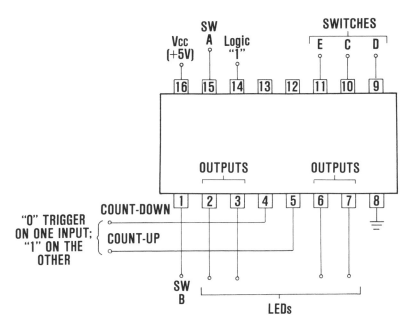

Figure 6-7: Up/down binary counter wiring diagram for Experiment 6-1

Typical clear, load, and count sequences are shown in Figure 6-8.

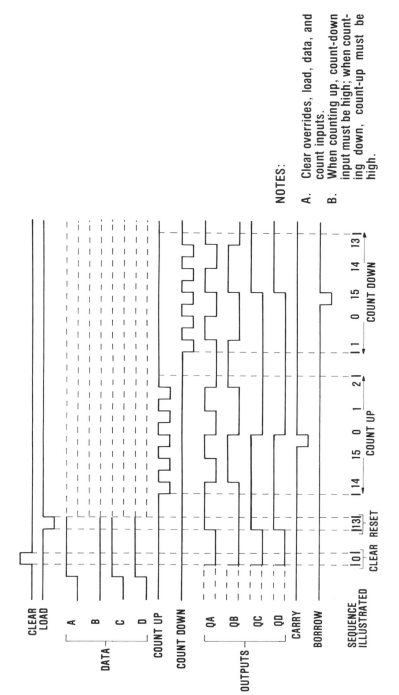

NOTES:

A. Clear overrides, load, data, and count inputs.
B. When counting up, count-down input must be high; when counting down, count-up must be high.

Figure 6-8: Typical clear, load, and count sequences for the binary counter 74193 (Courtesy Signetics Corp.)

Understanding Digital Circuits
Used for Register Operations

In keeping with the theme of this book — learning digital electronics on a budget — we will use the inexpensive 7495 4-bit right-shift, left-shift register to explain the digital circuits used to perform most of the basic operations required of a register.

As you have seen, a 74 TTL series register is a group of flip-flops used for temporary storage, and the number of flip-flops determines the amount of data per unit. Therefore, it follows, the 7495 4-bit shift register is constructed using four flip-flops. The circuit layout consists of four **R-S** master-slave flip-flops, four **AND-OR** invert gates, and six inverters configured to form a register that will perform right-shift, left-shift, or parallel-out operations, depending on what level is used on the input to the mode control.

The mode control input (pin 6) is located in the upper left-hand corner of the logic diagram (Figure 6-9). Right-shift operations are performed when you apply a logic level 0 to this mode control pin. Serial data is entered at the serial input (pin 1) and shifted one position right on each clock 1 pulse. When you are operating the register in the serial mode, clock 2 (pin 8) and parallel inputs A through D (pins 2, 3, 4, and 5) are inhibited. Each flip-flop and the **AND-OR** invert combinations connected to the **R-S** inputs are shown in Figure 6-9.

Data can be entered at the parallel inputs (pins 2, 3, 4, and 5) when a logical 1 level is applied to the mode control. The data entered at the parallel inputs is transferred to the data outputs (pins 10, 11, 12, and 13) on each clock 2 (pin 8) pulse. When you are operating in this mode, you can perform shift-left operations by externally tying the output of each flip-flop to the parallel input of the previous flip-flop (up to flip-flop C, with serial data entry at input D). *Note:* Information must be present at the R-S inputs prior to clocking, and transfer of data occurs on the falling edge of the clock pulse.

Now, let's say you are operating the **IC** in a shift-right mode, serial input/serial output. That is, the output of one flip-flop feeds into the inputs of the next. Table 6-4 shows the serial loading and transfer action of the register. This table shows the loading sequence and that the register is completely loaded on the fourth clock pulse. On the last clock pulse, the register contains the binary number 1010

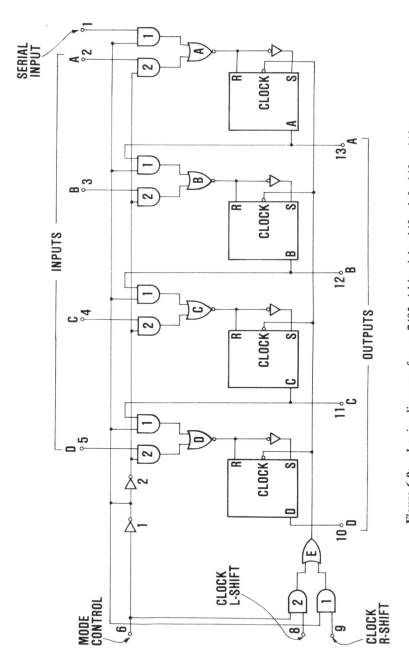

Figure 6-9: Logic diagram for a 7495 4-bit right-shift, left-shift, shift register (Courtesy Signetics Corp.)

in memory. It should be obvious that if you continue applying pulses, the data (binary number 1010) you have in the register will shift right on out of the **IC**. This means that the data will be completely lost unless you connect to another register or some other type digital circuit.

CLOCK 1 PULSE (R-SHIFT)	SERIAL INPUT	FLIP-FLOP 1	FLIP-FLOP 2	FLIP-FLOP 3	FLIP-FLOP 4
1	0	0	0	0	0
2	1	1	0	0	0
3	0	0	1	0	0
4	1	1	0	1	0

Table 6-4:   4-bit shift register operating in a right-shift mode

Although data can be moved from one register to another by serial shifting, this can be a slow process, depending on the time it takes to shift a certain number of bits and the clock frequency. As has been explained in previous pages, a faster method is to use a parallel transfer between registers. Another **IC** that you can use to accomplish parallel transfer is shown in Figure 2-2.

EXPERIMENT 6-2

Wiring and Testing a 4-Bit Right-Shift,

Left-Shift Register

In this experiment, two **IC**'s that you are already familiar with (a 7407, Project 3-1, and a 555 timer, Experiments 5-3 and 5-4) will be used with a 74194 4-bit bidirectional shift register. Figure 6-10 shows how the 74194 can be breadboarded to a set of control switches and to the 555 timer for a clock signal.

Referring to Chapter 4, Figure 4-18, you'll find that the 555 timer wiring diagrams shown in Figure 6-10 and 4-18 are basically the same. However, in this case, the clock pulse is manually generated

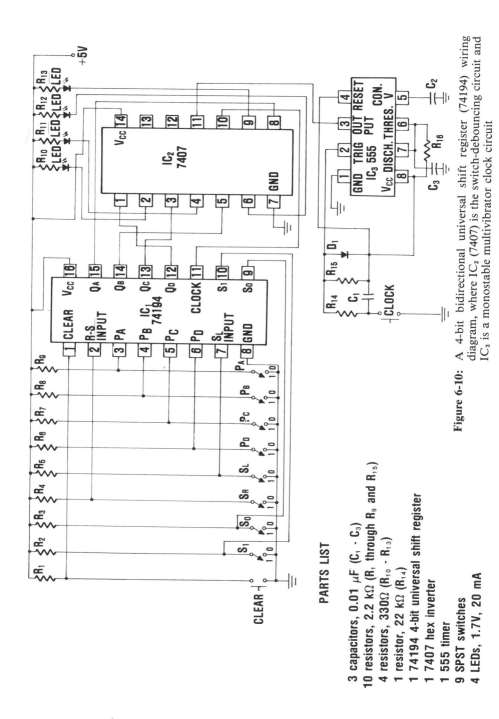

Figure 6-10: A 4-bit bidirectional universal shift register (74194) wiring diagram, where IC₂ (7407) is the switch-debouncing circuit and IC₃ is a monostable multivibrator clock circuit

PARTS LIST

3 capacitors, 0.01 μF (C₁ - C₃)
10 resistors, 2.2 kΩ (R₁ through R₉ and R₁₅)
4 resistors, 330Ω (R₁₀ - R₁₃)
1 resistor, 22 kΩ (R₁₄)
1 74194 4-bit universal shift register
1 7407 hex inverter
1 555 timer
9 SPST switches
4 LEDs, 1.7V, 20 mA

each time you depress the clock switch. When you activate the switch, you should find a *positive* pulse at the output (pin 3) of the 555 **IC**. But note, pin 10, the output of the inverter, should have an inverted pulse going into the clock input of the 74194 **IC**.

The 555 timer circuit has an 11 millisecond delay, which insures that only one pulse will be delivered to the shift register each time you activate the clock switch, even though there may be several bounces of the switch contacts. For this experiment, you do not have to worry about debouncing the other switches because you can get away with some switch contact bounce.

As shown (all switches open), the inputs are all at a logical high. When you close any of the switches, you are setting that input to a logical low. The outputs from the shift register are active-high, but to use LEDs on the outputs, you need active-low logic, which is why you must use the inverter stages of the 7404 hex inverter on the outputs of the 74194 register. If you are checking the circuit, you should find that the LEDs will light up whenever the outputs (QA, QB, QC, and QD) of the 74194 are at a logical high. The register has four distinct modes of operation. These are shown in Table 6-5.

	MODE S_1	CONTROL S_0
PARALLEL (BROADSIDE) LOAD	H	H
SHIFT RIGHT (IN THE DIRECTION QA TOWARD QD)	L	H
SHIFT LEFT (IN THE DIRECTION QD TOWARD QA)	H	L
INHIBIT CLOCK (HOLD)	L	L

Table 6-5: Four modes of operation that you can utilize when working with the 74194 shift register.

Parallel - In/Out Mode (the **IC** acts as a memory . . . stores a 4-bit word):

Step 1: Clear the register by depressing the clear switch.

Step 2: As shown in Table 6-5, you must set switches S_0 and S_1 to a logical high.

Step 3: Enter a 4-bit word (0101, or whatever you wish) by setting the switches on the parallel input pins (pins 3, 4, 5, and 6) of the register.

Step 4: Now, because in the parallel mode, data is loaded into the associated flip-flop and appears at the outputs only after a positive transition of the clock input, you will have to momentarily depress the clock switch. When you do this, you should see whatever 4-bit word you entered in Step 3, appear at the outputs.

Step 5: Check the ouputs (LEDs). There should be no change. The purpose of this step is to prove to yourself that the register will hold the first-entered data until you again depress the clock switch. In other words, the memory circuit will hold the last information entered until it is clocked again. *Note:* It is not necessary to clear the register before clocking in a new word. You should find that the circuit will write over the 4-bit word you previously entered. Also, during loading, the serial data flow is inhibited.

RIGHT SHIFT

Step 1: As shown in Table 6-5, set S_1 to low, S_0 to high.

Step 2: Depress the clear switch. This should set all LEDs to 0s.

Step 3: Set switch SR_1 (to pin 2 of the register **IC**) to either high or low.

Step 4: Activate the clear switch *one time* to enter that bit (see Step 3) into the QA section of the register **IC**.

Step 5: Enter one more bit at pin 2 (SR_1). Again activate the clear switch one time.

Step 6: Repeat Step 5 two more times, i.e., until you have four bits of data entered into the register.

Step 7: After you have completed the preceding six steps, you should have a 4-bit word entered and stored in the register, with the bit you first entered shown in the LED connected to the QD output of the register **IC**.

SHIFT LEFT

Refer to Table 6-5 and set mode control S_1 to high, S_0 to low. Now, simply repeat each step given in the procedure for the shift-right *except* you must now enter the new data in the shift-left serial input (pin 7) by using switch SL_1. Finally, you will find that it is impossible (or it should be) to clock the flip-flops in the register if both mode control inputs are at a logic 0 (switch closed). Incidentally, if you are having trouble with the experiment, check the mode control switches to see that they are not both connected to ground (not set to logical 0).

EXPERIMENT 6-3
Working with a Modulo Counter

A counter that is used to count specific amounts may be referred to as a *modulo* counter. In the previous pages, you have seen several **IC** counters that will meet this requirement; for example, the 7493 and 7490. The 7490 **IC** can be used for a modulo counter of from 2 to 6, with the use of R_{o1} and R_{o2} inputs, as was explained in detail in the first section of this chapter. Constructing a modulo counter using the 7493 **IC** is just as simple. First, wire the **IC** as shown in Figure 6-5 but do not ground pins 2 and 3 (R_{o1} and R_{o2}). To make the connections for a modulo-3 counter, simply complete the following steps.

Step 1: Connect R_{o1} (pin 2) to A (pin 12) and pin 1 (input B) to A.

Step 2: Short pins 9 and 3 (R_{o2} and B).

Figure 6-11 shows the wiring diagram for Steps 1 and 2.

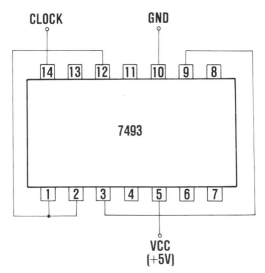

Figure 6-11: Pin connection diagram for a modulo-3 counter using a 7493 IC (see Figure 6-5 and text for additional wiring instructions)

To construct a modulo-7, 11, 13, 14, and 15 counter, you will need an additional **IC**. A 7408 quadruple 2-input positive **AND** gate (see Figure 2-8) will fill this requirement. Figure 6-12 shows the wiring diagram for modulo-7 counter constructed using a 7493 4-bit binary counter and a 7408 **AND** gate.

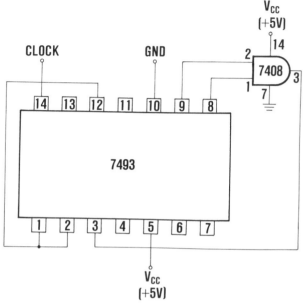

Figure 6-12: Modulo-7 counter constructed using a 7493 4-bit binary counter and a 7408 AND gate IC

MODULO COUNTER	CONNECTIONS RO(1)	CONNECTIONS RO(2)	COMMENTS
7	A	BC	7408 FOR B AND C
11	A	BD	7408 FOR B AND D
13	A	CD	7408 FOR C AND D
14	B	CD	7408 FOR C AND D
15	AB	CD	7408 FOR A AND B AND C AND D

Table 6-6: Modulo counter wiring connections for the schematic diagram shown in Figure 6-5, when using a 7408 AND gate with the 7493 IC

To wire the circuits for the other counters requiring an **AND** gate, simply follow the instructions given in Table 6-6. Several other modulo counters (4, 5, 6, 8, 9, and 10) can be constructed if you modify the circuit, as shown in Table 6-7.

MODULO COUNTER	CONNECTIONS		COMMENTS
	RO1	RO2	
4	C	C	TIE TOGETHER R_{01} AND R_{02}
5	A	C	——
6	B	C	——
8	D	D	TIE TOGETHER R_{01} AND R_{02}
9	A	D	——
10	D	B	——

Table 6-7: Six additional modulo counters that can be constructed when using a 7493 IC wired as shown in Figure 5-6

CHAPTER 7

Practical Applications Using
Digital Arithmetic IC's

The experiments in this chapter will give you the necessary "hands-on" experience that will help you troubleshoot digital circuits faster and more efficiently. Each example and experiment in the following pages will use a **TTL** device, because, for the experimenter, a **TTL** or **CMOS IC** is the most economical to work with. Incidentally, the 7400 series is pin-for-pin compatible with 74C-**CMOS IC**'s, meaning that you can transfer the knowledge you gain from these experiments to any similar digital circuits using the newer **CMOS** devices.

To show you what we mean by that last statement, take, for example, the 7402 TTL quadruple 2-input **NOR** gate shown in Figure 2-15. This IC is pin-for-pin compatible with the CD 4001 (a **CMOS**) quad 2-input **NOR** gate. Which means, as we have said, learn about one family (**TTL**) and you have a fundamental knowledge of the other (**CMOS**).

This chapter will help you "short-circuit" the time normally involved in gaining the experience needed to work with arithmetic logic units found in virtually all microcomputer systems. There also are practical aids for arithmetic operations using **IC**'s, as well as logic

176

functions, that you can use to acquire the skill needed to work with today's digital circuits.

How Digital IC's
Perform Addition

In Chapter 2, it was stated that an adder is an arrangement of logic gates that add two binary digits and produce sum and inverted carry outputs. Figure 7-1 shows a truth table and a logic diagram of the simplest arrangement of logic gates that one can use to add two 1-bit numbers. The circuit adds two 1-bit numbers (A and B) and yields a sum (S) and a carry bit (C).

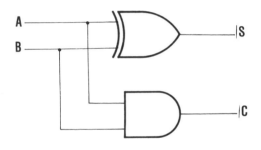

INPUTS		OUTPUTS	
A	B	S	C
0	0	0	0
0	1	1	0
1	0	1	0
1	1	0	1

$$S = A\bar{B} + \bar{A}B$$
$$C = AB$$

Figure 7-1:   Truth table and half-adder constructed using exlusive OR/AND gates

Referring to the top line of the truth table in Figure 7-1, you will note that if inputs A and B are 0, outputs S and C will be 0. In other words, 0 plus 0 equals a sum 0 and a carry of 0. Note, this is exactly in line with what you do when following the rules of ordinary decimal arithmetic. The next two lines of the truth table show that the sum of 1 plus 0 is equal to 1, no matter whether you enter 1 into input A or B. The carry in these two operations, however, is still 0. Moving down to the last line, adding 1 and 1 yields a sum of 0 and a carry of 1. This point is where the rules of binary addition given in Chapter 2 come into play. For example, let's say that we are adding binary 1 and 1. Referring to Table 2-1, we see that in this instance, 01 plus 01 equals 10. Or, converting to decimal equivalents, 1 plus 1 equals 2.

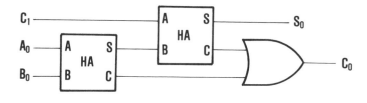

C_i	B	A	S_0	C_0
0	0	0	0	0
0	0	1	1	0
0	1	0	1	0
0	1	1	0	1
1	0	0	1	0
1	0	1	0	1
1	1	0	0	1
1	1	1	1	1

Figure 7-2:   A basic full-adder constructed using two half-adders (HA)

As we can quickly see, this type adder (called a *half-adder*) is very limited. It can be cascaded with other adder circuits to handle numbers of any desired length, but you are still limited to adding the two least significant bits because there are no provisions for entering a carry back into the adder for more than the 2-bit column of numbers. Is there a way to beat this problem? Yes. Simply combine two half-adders with an **OR** function to create what is called a *full-adder*. The full-adder has three inputs; i.e., it sums three bits at one time, an A bit, a B bit and a carry-in bit (C_i) from the previous stage. Also, it has two outputs, S_o (sum out) and C_o (carry out). See Figure 7-2.

Essentially, the truth table in Figure 7-2 shows that the S_o output will be a 1 if only one input is high (see the second and third lines), the C_o output will be a 1 if any two inputs are high (see lines 4, 6, 7, and 8), and both S_o and C_o outputs will be 1 if all inputs are high (see the bottom line).

In Chapter 1, we discussed a single-bit, binary adder (the 7480). However, in most digital work, a 4-bit binary full-adder is the most useful. The difference between the single-bit adder and the parallel addition adder (for example, the 7483), so far as hardware is concerned, is that parallel addition requires a full-adder for each bit of the binary number. When the 7483 **IC** is referred to as a 4-bit full-adder, it is understood that it is being used for parallel addition. But, as we have said, the same **IC** can be wired to produce two single-bit full-adders for serial addition. More about these operations a little later. Figure 7-3 shows a truth table and pin configuration for a 16-lead dual in-line, molded package, 7483 IC.

If it is assumed, for the moment, that you want the IC to be used in a parallel operation, the carry-in terminal is fixed at logic 0 (grounded), and the **IC** will sum any two 4-bit binary numbers you apply to the A and B inputs. The sum of the two numbers appears at output Σ_1 through Σ_4, while the carry-out bit appears at C_4. As is stated in the notes in Figure 7-3, C_2 is an internal carry value, which means that the user does not have access to these internally connected carry bits.

To permit summing of more than two 4-bit numbers, it is common practice to cascade two or more of these adder **IC**'s. Typically, you will find summing of 8-, 12-, and 16-bit numbers when the cascading process is used. The cascading procedure is very simple. Connect the carry-out pin (14) labeled C_4, of one **IC** to the carry-

Pin configuration, 7483:

Pin	Signal	Pin	Signal
1	A₄	16	B₄
2	Σ₃	15	Σ₄
3	A₃	14	C₄
4	B₃	13	C₀
5	VCC	12	GND
6	Σ₂	11	B₁
7	B₂	10	A₁
8	A₂	9	Σ₁

NOTES:

Input conditions at A_1, A_2, B_1, B_2, and C_0 are used to determine outputs Σ_1 and Σ_2, and the value of the internal carry C_2. The values at C_2, A_3, B_3, A_4, and B_4 are then used to determine outputs Σ_3, Σ_4, and C_4.

Truth table:

INPUT				OUTPUT					
				WHEN $C_0=0$			WHEN $C_0=1$		
A_1	A_2	B_1	B_2	Σ_1	Σ_2	C_2	Σ_1	Σ_2	C_2
A_3	A_4	B_3	B_4	Σ_3	Σ_4	C_4	Σ_3	Σ_4	C_4
0	0	0	0	0	0	0	1	0	0
1	0	0	0	1	0	0	0	1	0
0	0	1	0	1	0	0	0	1	0
1	0	1	0	0	1	0	1	1	0
0	1	0	0	0	1	0	1	1	0
1	1	0	0	1	1	0	0	0	1
0	1	1	0	1	1	0	0	0	1
1	1	1	0	0	0	1	1	0	1
0	0	0	1	0	1	0	1	1	0
1	0	0	1	1	1	0	0	0	1
0	0	1	1	1	1	0	0	0	1
1	0	1	1	0	0	1	1	0	1
0	1	0	1	0	0	1	1	0	1
1	1	0	1	1	0	1	0	1	1
0	1	1	1	1	0	1	0	1	1
1	1	1	1	0	1	1	1	1	1

Figure 7-3: Truth table and pin configuration for a 16-lead dual in-line, molded package, 7483 IC

in pin (13) labeled C_o, of the next **IC**. The **IC** handling the four lower order bits should have its carry-input pin grounded (at logic 0). The carry-out pin of the last adder can be used to read the sum, or as an overflow detector.

How to Subtract Binary Numbers

As with addition, subtraction of numbers in digital equipment is performed with numbers expressed in binary form. When subtracting binary numbers, there are only four conditions that will be met. These are:

$$
\begin{array}{cccc}
0 & 1 & 1 & 0 \\
-0 & -0 & -1 & -1 \\
\hline
0 & 1 & 0 & 1 \text{ and borrow}
\end{array}
$$

From the last basic subtraction rules, we find that subtracting binary 1 from 0 requires a borrow of 1 from the next most significant bit. When 1 is borrowed, it is brought back as 1 plus 1 and, subtracting, we get 1. This result is expressed as 1 and a borrow of 1. If this sounds confusing, don't be alarmed. It is probably the most bewildering rule in binary arithmetic. To help clarify the rules, subtract decimal − 3 from + 9, using binary arithmetic. First, using Table 2-1, convert the decimal 3 and 9 to binary. Your answer is, 3 decimal = 0011 and 9 decimal = 1001. Next, set up the example problem.

$$
\begin{array}{r}
1001 \\
-0011 \\
\hline
\end{array}
$$

Then subtract, using the four basic subtraction rules.

$$
\begin{array}{ll}
\quad\text{011 (borrowed 1)} & \\
\text{binary } 1\emptyset\emptyset1 & \quad\text{decimal} \quad 9 \\
\quad -0011 \quad \text{equals} & \quad\quad\quad\quad -3 \\
\hline
\quad\quad 110 & \quad\quad\quad\quad\quad 6
\end{array}
$$

Now, starting at the right, notice that the first bits (1 and 1) produce an answer of 0. However, the next (second from right) requires a borrow 1. Why? Because we cannot subtract a 1 from a 0 without a borrow 1. But, notice that the borrowed number must be taken from the column at the extreme left. Figure 7-4 shows a logic diagram that could perform 1-bit binary subtraction. The circuit illustrated is called a *half-subtractor circuit*.

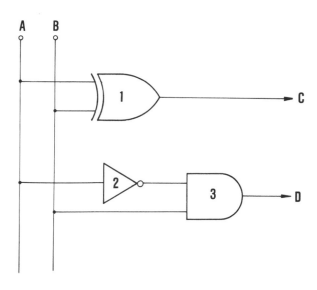

A	B	C	D
0	0	0	0
0	1	1	1
1	0	1	0
1	1	0	0

Figure 7-4:   Half-subtractor logic diagram with truth table

Looking at the truth table and logic diagram in Figure 7-4, you can see there are two inputs (labeled A and B) and two outputs, which are the *difference* and *borrow* (labeled C and D). In Chapter 2, you studied an exclusive **OR** gate and its Boolean equation. All we have here, so far as the top logic symbol is concerned, is an exclusive **OR** gate. Therefore, the difference expression is the exclusive **OR** function of Inputs A and B. The Boolean equation can be written for this output as:

$$C = \overline{A} B + A \overline{B}$$

On the other hand, borrow (B) output is the **AND** function of the complement of input A with input B. The Boolean equation for the D output is:

$$D = \overline{A} B$$

We have shown that digital circuits normally involve more than one bit, therefore, it stands to reason that to subtract a 4-bit binary number requires additional circuits. For this type operation, a full-subtractor, in addition to the half-subtractor circuit, is needed. However, since we have explained the rules for binary arithemetic and how to subtract a 4-bit binary number, we shall now look at another method of subtracting that uses less digital circuitry.

Using Digital IC's to Subtract

The 7483 is a 4-bit binary full-adder and usually is used in parallel addition, as has been discussed. But this same **IC** can be used to subtract by using *complement* methods. With complements, we can find the difference of two binary numbers by the addition process instead of directly through subtraction. There are two complement methods that may be used with the 7483. These are the *1's complement* and *2's complement*.

If you wish to form the 1's complement of a binary number, it's easy. All you need to do is change all 1's in the number to 0's and all 0's to 1's. For example, the 1's complement of the binary number 1010 is 0101. An inverter such as the 7404 hex inverter shown in Figure 3-17 will do this job quite nicely. Just feed the binary number

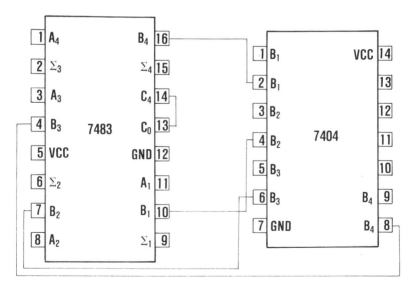

Figure 7-5: 1's complement subtraction circuit using a 7483 IC and a 7404 inverter

in and you will get the complement out. Figure 7-5 shows how a 7483 can be connected with inverters to form a 1's complement subtraction circuit.

After the binary number is passed through the inverter, we have the complement. Then the complement of the binary number to be subtracted (subtrahend) is *added* to the number you are subtracting from (minuend), instead of subtracting. The 1's complement subtraction is completed by performing what is called an *end-around carry*. This is done by adding the carry from the most significant bit (MSB) in the sum to the least significant bit (LSB) of the sum, to get the difference. To get a better idea of how this is accomplished, look at the following example.

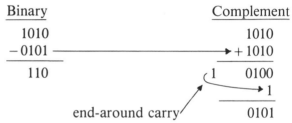

Now, let's see how the **IC** circuits shown in Figure 7-5 perform this task. First, as you can see, inverters produce the 1's complement of the subtrahend (0101 in the example just given), inputs B_4, B_3, B_2, and B_1 of the 7407. The complements (1010) of the 7407 are labeled $\overline{B}_4$, $\overline{B}_3$, $\overline{B}_2$, and $\overline{B}_1$. Next, the 7483 4-bit binary adder does the job of producing the sum of the inputs at A_4, A_3, A_2, and A_1; i.e., the minuend 1010 and 1's complement of 1010.

In this process, the carry output of the adder is sent back to the adder carry input (pin 13 shorted to pin 14). This return connection is your end-around carry operation. Taking A_1 and $\overline{B}_1$ inputs as an example, after a carry-out bit is sent back to the carry-in, it is once more added to these inputs to give the difference. The remaining bits (the A's and $\overline{B}$'s) are again added to give the final difference. The outputs of the adder are Σ_4, Σ_3, Σ_2, and Σ_1. The entire process using the 1's complement subtraction circuit is:

$$\text{inputs: } A_4, A_3, A_2, A_1 = 1010$$
$$B_4, B_3, B_2, B_1 = 0101$$

$$\text{output of the inverters: } \overline{B}_4, \overline{B}_3, \overline{B}_2, \overline{B}_1 = 1010$$

combining $\overline{B}$ inputs to A inputs:

$$1010A$$
$$+\ 1010B$$
$$\overline{0100\Sigma}$$

The carry output is the most significant bit of the sum output.

$$\text{carry-out} = 1$$

The logic 1 is returned to the carry-in input of the adder IC and the addition is repeated. The result is:

$$1 \text{ carry-in}$$
$$1010A$$
$$+\ 1010B$$
$$\overline{0101\Sigma}$$

The carry-out bit of the sum is eliminated to give the difference by the 1's complement method.

The 2's complement method is a second form of complementing that you can use with the 7483. It requires the connections shown in Figure 7-6. Note, the wiring is similar to that in Figure 7-5 except for the carry input, which must (in this case) be held at a logic level 1 for subtraction.

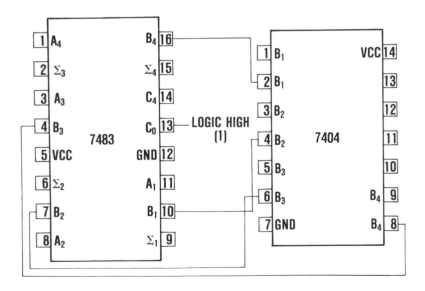

Figure 7-6: 2's complement method using a 7483 and inverter IC

As before, the inverters change the subtrahend inputs; i.e., the B's to their complement $\overline{B}$'s. The purpose of holding the carry input at a logic high is so the **IC** can add this extra 1 to the complemented number. The following example will show how this method performs subtraction.

decimal	normal binary	with the 7483 **IC**
15	1000	1 carry-in
-8	-1111	$+1111A$
7	0111	$\cancel{1}0111$
		eliminating the carry-in
		bit $= 0111$

EXPERIMENT 7-1
A Practical 4-Bit Magnitude Comparator

There are many applications where it is important to determine which one of the two numbers is greater or less than the other.

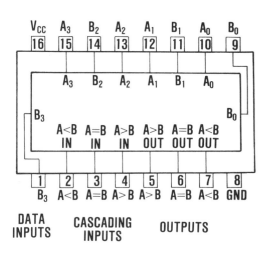

Figure 7-7:   Pin configuration of a 4-bit magnitude comparator 7485 IC

This is a job for the magnitude comparator **IC**. Figure 7-7 shows the pin configuration for a 7485 4-bit magnitude comparator. Stated simply, a magnitude comparator such as this one compares the values of two binary numbers—typically designated A and B—and generates a logic 1 level at one of the three outputs. As you can see, the outputs indicate A = B, A > B, and A< B.

The 7485 4-bit comparator has four inputs for each short term storage circuit (register A and register B, and three outputs). These outputs indicate which A binary data is larger than B binary data, if A is less than B, or if A is equal to B. If we ignore the cascading inputs (pins 2, 3, and 4) for the moment, we can look at the truth table for the basic 4-bit magnitude comparator functions (see Table 7-1).

COMPARING INPUTS				OUTPUTS		
A_3, B_3	A_2, B_2	A_1, B_1	A_0, B_0	A>B	A<B	A=B
$A_3>B_3$	X	X	X	H	L	L
$A_3<B_3$	X	X	X	L	H	L
$A_3=B_3$	$A_2>B_2$	X	X	H	L	L
$A_3=B_3$	$A_2<B_2$	X	X	L	H	L
$A_3=B_3$	$A_2=B_2$	$A_1>B_1$	X	H	L	L
$A_3=B_3$	$A_2=B_2$	$A_1<B_1$	X	L	H	L
$A_3=B_3$	$A_2=B_2$	$A_1=B_1$	$A_0>B_0$	H	L	L
$A_3=B_3$	$A_2=B_2$	$A_1=B_1$	$A_0<B_0$	L	H	L
$A_3=B_3$	$A_2=B_2$	$A_1=B_1$	$A_0=B_0$	H	L	L
$A_3=B_3$	$A_2=B_2$	$A_1=B_1$	$A_0=B_0$	L	H	L
$A_3=B_3$	$A_2=B_2$	$A_1=B_1$	$A_0=B_0$	L	L	H

NOTE: H = HIGH LEVEL
L = LOW LEVEL
X = IRRELEVANT

Table 7-1: Truth table for a basic 4-bit magnitude comparator. When used in this configuration, the A = B input (pin 3) must be connected to logic 1 while the A < B (pin 2) and A > B (pin 4) inputs should be wired to a logic low.

The wiring diagram for this **IC** is shown in Figure 7-8. "A" word input switches can be connected to pins 10, 12, 13, and 15. "B" word input switches are connected to pins 9, 11, and 14. Next, connect LED's to pins 5, 6, and 7, as shown. Refer to Table 7-1 for switch input settings; that is, various combinations of A and B inputs.

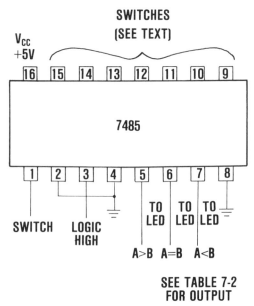

Figure 7-8: Wiring diagram for a basic 4-bit magnitude comparator using a 7485 IC

Notice, pin 3 of the cascading inputs must be placed in a constant logic 1 condition. If this is not done, the circuit cannot output the A = B condition. The other two cascading inputs (pins 2 and 4) are grounded. If you are cascading 7485's to build 8-, 12-, or 16-bit magnitude comparators, it is merely a matter of leaving the cascading inputs ungrounded and connecting the outputs of one stage to the corresponding cascading inputs of the next stage. Of course, pin 3 is not connected to a logic high either. It goes on to the next **IC**, as do pins 2 and 4. Table 7-2 can be used to test various combinations of binary numbers when cascading these **IC's**.

Typical applications where you'll find comparators are computer circuits, antenna control systems, and other industrial type servomechanisms. The computer circuits that use comparators are

COMPARING INPUTS				CASCADING INPUTS			OUTPUTS		
A_3, B_3	A_2, B_2	A_1, B_1	A_0, B_0	$A>B$	$A<B$	$A=B$	$A>B$	$A<B$	$A=B$
$A_3>B_3$	X	X	X	X	X	X	H	L	L
$A_3<B_3$	X	X	X	X	X	X	L	H	L
$A_3=B_3$	$A_2>B_2$	X	X	X	X	X	H	L	L
$A_3=B_3$	$A_2<B_2$	X	X	X	X	X	L	H	L
$A_3=B_3$	$A_2=B_2$	$A_1>B_1$	X	X	X	X	H	L	L
$A_3=B_3$	$A_2=B_2$	$A_1<B_1$	X	X	X	X	L	H	L
$A_3=B_3$	$A_2=B_2$	$A_1=B_1$	$A_0>B_0$	X	X	X	H	L	L
$A_3=B_3$	$A_2=B_2$	$A_1=B_1$	$A_0<B_0$	X	X	X	L	H	L
$A_3=B_3$	$A_2=B_2$	$A_1=B_1$	$A_0=B_0$	H	L	X	H	L	L
$A_3=B_3$	$A_2=B_2$	$A_1=B_1$	$A_0=B_0$	L	H	X	L	H	L
$A_3=B_3$	$A_2=B_2$	$A_1=B_1$	$A_0=B_0$	L	L	H	L	L	H

NOTE: H = HIGH LEVEL
L = LOW LEVEL
X = IRRELEVANT

Table 7-2: Function table for a 7485 that can be used when cascading these IC's to build 8-, 12-, or 16-bit magnitude comparators

usually a "go" or "no go" type operation. For example, if the numbers are equal, the computer performs a certain operation. If they are not, it may perform either of two other possibilities, depending on whether the A is greater or less than B.

In the case of servomechanisms, you'll generally find that a digital circuit monitors the parameter (such as antenna elevation or azimuth) to be controlled. This circuit, the control circuit, generates a binary number that usually is proportional to the actual antenna position, etc. The output, a binary number, is sent to the magnitude comparator, where it is compared to a reference (another binary number representing the desired antenna position). If the actual antenna position is above or below, to the right or left, etc., that magnitude comparator generates a "greater-than" or "less-than" output, which causes the servomechanism to try to correct for a zero difference. Antenna systems such as this are used for tracking spacecraft, airplanes, and other moving vehicles.

EXPERIMENT 7-2
Wiring Up and Checking a 4-Bit BCD Adder

By wiring the 7483 full-adder for normal operation and including a 7404 hex inverter, you are able to include LED's in the output circuits of the 7483. The sum of the two binary numbers will appear in an active-high format at the output pins 9, 6, 2, and 15. *Note:* Any sum that exceeds the binary number 1111 (decimal 15) will cause the LED connected to the carry-out pin of the 7483 to light up, indicating an error or overflow. However, as we stated earlier, it is possible to use the carry-out data to read larger binary number sums. . . up to 11110, where the left-hand bit is the carry output. To see how you can add up to a sum of 11110, simply assume that you want to add the decimal number 15 plus 15. To do this, set all eight input switches to a logic high. Your LED's (all five) should light up except one, giving you an answer of 11110, or the decimal number 30.

You can use the circuit shown in Figure 7-9 to perform this experiment. The purpose of the 7404 hex inverter is to change the active-high outputs of the 7483 to active-low inputs, which is required to drive the cathodes of the LED's. See Project 3-1 for LED current-limiting resistor value calculations.

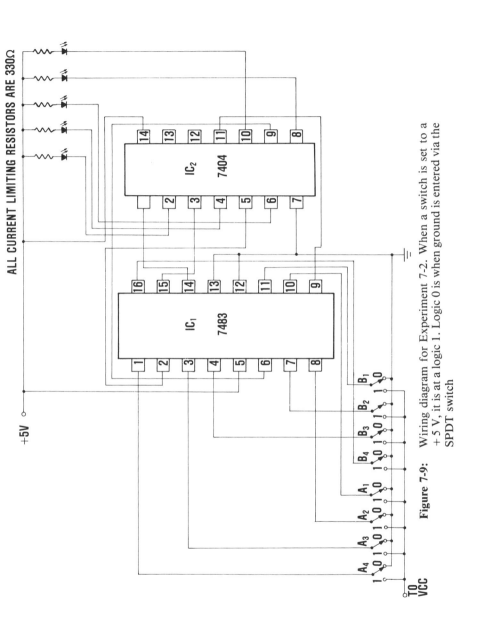

Figure 7-9: Wiring diagram for Experiment 7-2. When a switch is set to a +5 V, it is at a logic 1. Logic 0 is when ground is entered via the SPDT switch

A Guide to Arithmetic Logic Units

At one time, just about any electronic calculating device you opened up contained a 74181 arithmetic logic unit (ALU). But then, companies such as National Semiconductor introduced a series of special function single-chip calculator IC's, together with a compatible program chip, capable of converting any of the calculator IC's into a fully programmable "learn mode" calculator. Some of National's first IC's were the MM5760 slide rule circuit, the MM5762 business and financial calculator, the MM5764 international conversion calculator, and the MM5765 calculator program. If one does not have a thorough understanding of computer-oriented logic and arithmetic operations, it is rather difficult to appreciate the real power of today's offsprings of these IC's. Since sophisticated operations such as log and trig functions are beyond the scope of this book, we will limit our investigations to the more elementary operations of the 74181 and 74S381. Figure 7-10 shows the pin configuration of the 74181.

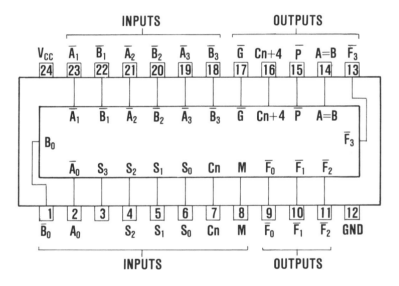

Figure 7-10: The 74181 ALU IC pin configuration

This IC accepts two sets of 4-bit binary numbers (words), labeled as A's and B's. These inputs are treated much like the inputs to the 4-bit adder IC you previously studied. This system performs 16 binary arithmetic operations on two 4-bit words (as shown in Table 7-3A and B), depending on the 4-bit code applied to the selected inputs S_0 through S_3.

A

SELECTION $S_3\,S_2\,S_1\,S_0$	M = H LOGIC FUNCTION	ACTIVE-HIGH DATA — M = L; ARITHMETIC OPERATIONS $C_n=0$ ($\overline{C}_n=1=H$)	$C_n=1$ ($\overline{C}_n=0=L$)
L L L L	$F=\overline{A}$	$F=A$	$F=A$ PLUS 1
L L L L	$F=\overline{A+B}$	$F=A+B$	$F=(A+B)$ PLUS 1
L L H L	$F=\overline{A}\,B$	$F=A+\overline{B}$	$F=(A+\overline{B})$ PLUS 1
L L H H	$F=0$	$F=$ MINUS 1 (2's COMPL)	$F=$ ZERO
L H L L	$F=A\overline{B}$	$F=A$ PLUS $A\overline{B}$	$F=A$ PLUS $A\overline{B}$ PLUS 1
L H L H	$F=\overline{B}$	$F=(A+B)$ PLUS $A\overline{B}$	$F=(A+B)$ PLUS $A\overline{B}$ PLUS 1
L H H L	$F=A\oplus B$	$F=A$ MINUS B MINUS 1	$F=A$ MINUS B
L H H H	$F=A\overline{B}$	$F=A\overline{B}$ MINUS 1	$F=A\overline{B}$
H L L L	$F=\overline{A}+B$	$F=A$ PLUS AB	$F=A$ PLUS AB PLUS 1
H L L H	$F=\overline{A\oplus B}$	$F=A$ PLUS B	$F=A$ PLUS B PLUS 1
H L H L	$F=B$	$F=(A+\overline{B})$ PLUS AB	$F=(A+\overline{B})$ PLUS AB PLUS 1
H L H H	$F=AB$	$F=AB$ MINUS 1	$F=AB$
H H L L	$F=1$	$F=A$ PLUS A	$F=A$ PLUS A PLUS 1
H H L H	$F=A+\overline{B}$	$F=(A+B)$ PLUS A	$F=(A+\overline{B})$ PLUS A PLUS 1
H H H L	$F=A+B$	$F=(A+\overline{B})$ PLUS A	$F=(A+B)$ PLUS A PLUS 1
H H H H	$F=A$	$F=A$ MINUS 1	$F=A$

B

SELECTION $S_3\,S_2\,S_1\,S_0$	M = H LOGIC FUNCTION	ACTIVE-LOW DATA — M = L; ARITHMETIC OPERATIONS $C_n=0$ ($C_n=0=L$)	$C_n=1$ ($C_n=1=H$)
L L L L	$F=\overline{A}$	$F=A$ MINUS 1	$F=A$
L L L H	$F=\overline{AB}$	$F=AB$ MINUS 1	$F=AB$
L L H L	$F=\overline{A}+B$	$F=A\overline{B}$ MINUS 1	$F=A\overline{B}$
L L H H	$F=1$	$F=$ MINUS 1 (2's COMPL)	$F=$ ZERO
L H L L	$F=\overline{A+B}$	$F=A$ PLUS $(A+\overline{B})$	$F=A$ PLUS $(A+\overline{B})$ PLUS 1
L H L H	$F=\overline{B}$	$F=AB$ PLUS $(A+\overline{B})$	$F=AB$ PLUS $(A+\overline{B})$ PLUS 1
L H H L	$F=\overline{A\oplus B}$	$F=A$ MINUS B MINUS 1	$F=A$ MINUS B
L H H H	$F=A+\overline{B}$	$F=A+\overline{B}$	$F=(A+\overline{B})$ PLUS 1
H L L L	$F=\overline{A}B$	$F=A$ PLUS $(A+B)$	$F=A$ PLUS $(A+B)$ PLUS 1
H L L H	$F=A\oplus B$	$F=A$ PLUS B	$F=A$ PLUS B PLUS 1
H L H L	$F=B$	$F=A\overline{B}$ PLUS $(A+B)$	$F=A\overline{B}$ PLUS $(A+B)$ PLUS 1
H L H H	$F=A+B$	$F=A+B$	$F=(A+B)$ PLUS 1
H H L L	$F=0$	$F=A$ PLUS A	$F=A$ PLUS A PLUS 1
H H L H	$F=A\overline{B}$	$F=AB$ PLUS A	$F=AB$ PLUS A PLUS 1
H H H L	$F=AB$	$F=A\overline{B}$ PLUS A	$F=A\overline{B}$ PLUS A PLUS 1
H H H H	$F=A$	$F=A$	$F=A$ PLUS 1

Table 7-3:   Function tables for the 74181 ALU (Courtesy Signetics Corp.)

The various functions controlled by the select inputs include addition, subtraction, decrement, and straight transfer. When performing arithmetic manipulations, the internal carries must be enabled by applying a low-level voltage to the mode control input (M). The primary outputs of the IC are the F pins. For example, if two 4-bit binary numbers A and B are to be summed, the result would appear at the F outputs. Table 7-3 shows two function tables, active-high and active-low. This IC will accommodate either one of these data inputs if pin designations are re-interpreted as shown in Table 7-4.

PIN NUMBER	2	1	23	22	21	20	19	18	9	10	11	13	7	16	15	17
ACTIVE-HIGH DATA	A_0	B_0	A_1	B_1	A_2	B_2	A_3	B_3	F_0	F_1	F_2	F_3	C_n	C_{n+4}	X	Y
ACTIVE-LOW DATA	$\overline{A_0}$	$\overline{B_0}$	$\overline{A_1}$	$\overline{B_1}$	$\overline{A_2}$	$\overline{B_2}$	$\overline{A_3}$	$\overline{B_3}$	$\overline{F_0}$	$\overline{F_1}$	$\overline{F_2}$	$\overline{F_3}$	C_n	C_{n+4}	$\overline{P}$	$\overline{G}$

Table 7-4:   Pin designations for active-high or active-low data format

Referring to Figure 7-10, the A = B output (pin 14) is simply a special equality function output. Three outputs (pins 15, 16, and 17) are carry circuits. They are used primarily when cascading to other IC's. The basic idea when using pins 15 and 17, is to reduce the time required for arithmetic operations. For example, the typical addition time for the 74181 is 24 nanoseconds for 4-bits. When expanding (cascading), and used with a look-ahead carry generator (74182) for 16-bit addition, only 13 nanoseconds further delay is added, so that the total time for addition is only 37 nanoseconds instead of 48 nanoseconds. If high speed is not of importance, a ripple-carry input (C_n) and a ripple-carry output ($C_n + 4$) are available, as shown in Figure 7-10 (see pins 7 and 16).

The two function tables summarize the logic and arithmetic operations possible with this IC. By referring to the top portion of the tables, you will note that the operations performed at any given moment depend on what 4-bit binary word you apply to the S inputs (S_0 through S_3) and what logic signal (high or low) you apply to the M and C_n inputs, where L = low (0) and H = high (1).

To test an operation of the system, first M is fixed at logic H. You get this information from the first column of logic functions (active-high data) in Table 7-3. Now, if you set the selection inputs (S_0, S_1, S_2 and S_3) to 0000 (LLLL), the table shows that F = $\overline{A}$. In other words, the F outputs are an inverted version of the A inputs. Or, to put it another way, F is the 1's complement of A.

What if you want F to equal $A\overline{B}$? To answer this question, simply look at the table under the selections inputs and the eighth line down shows that LHHH (0111) on these inputs will produce $F = A\overline{B}$. The last line of this truth table shows that if you set the selection inputs to HHHH, F = A; i.e., the F outputs are identical to the A inputs.

Let's stop here and take a closer look at what is really happening when you set the selection inputs to a binary number such as HLHH (F = AB). First of all, you'll recognize this logic formula as that of an **AND** gate. But what we are talking about, in this case, are two different 4-bit binary words rather than a pair of 1-bit logic levels. When working with **IC's** such as this one, it must be remembered that F = AB means that the corresponding elements of each 4-bit word are **AND**ed together: $F_0 = A_0 \bullet B_0$, $F_1 = A_1 \bullet B_1$, $F_2 = A_2 \bullet B_2$, and $F_3 = A_3 \bullet B_3$. Also, you must realize that this same general idea applies to all other logic functions listed in the M = H column of Table 7-3.

Now that we have a general idea of the logic function column (M = H), let's direct our attention over to the M = L column and see what other operations are possible with this **IC**. Referring to Table 7-3, active-high data, you will see that there are arithmetic functions, logic functions, and, in some cases, combinations of logic and arithmetic functions. Notice, the manufacturer has spelled out in plain words all arithmetic operations. You will find that the word *plus* has been used to mean the arithmetic operation of addition. But the + operator is used when it is a *logic* **OR** operation. Just remember these two rules and you should not encounter any confusion when reading the M = L arithmetic operations. However, as you can see, there are two different columns under this heading, $\overline{C}_n = O = L$ and $\overline{C}_n = 1 = H$.

To understand this operation, let's first consider the $\overline{C}_n = 1$ column. For instance, if you want to add two binary numbers (A and B), the tenth line shows F = A plus B when M = H, and the select inputs are set at HLLH. In this setup, the F outputs will be equal to the arithmetic sum of inputs A and B. Again, referring to the same column, the first three lines are clearly logic operations: $F = \overline{A}$, F = A + B, and $F = A + \overline{B}$.

In previous sections of this chapter, we have said quite a bit about 2's complement. The fourth line of the M = 0, $\overline{C}_n = 1$ and where the selection inputs are set at LLHH, the F output generates the 2's complement form of -1.

The 74181 can also be used as a comparator. The $A = B$ output is internally decoded from the function ouputs (F_0, F_1, F_2, and F_3) so that when you apply two 4-bit binary numbers of equal magnitude to A and B inputs, it will assume a high-level state to indicate equality ($A = B$). The **IC** should be in the subtract mode (LHHL) when performing this comparison. Subtraction is accomplished by 1's complement addition where the 1's complement of the subtrahend is generated internally. The resultant output is $A = B = 1$, which requires an end-around or forced-carry to provide $A - B$.

The carry output ($C_n + 4$) can also be used to determine the relative magnitude of two numbers. Again, the **IC** should be placed in the subtract mode by placing the control lines at LHHL. In all, this **IC** performs 16 binary arithmetic operations on two 4-bit words, as shown in the Function table. This same ALU also provides 16 possible functions of two Boolean variables without the use of external circuitry. You select these logic functions, such as $F = A + B$, $F = A + \overline{B}$, etc., by use of the four function-select inputs (S_0, S_1, S_2, and S_3), with the mode control input (M) at a high logic level to disable the internal carry. You will find all 16 logic functions are

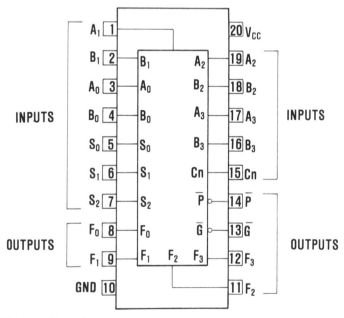

Figure 7-11: Arithmetic logic unit/function generator 74S381 pin configuration. Typical operating supply voltage (V_{cc}) is 5 V., max. low level input is 0.8 V., min. high level is 2 V. (Courtesy Monolithic Memories)

detailed in Table 7-3 and include exclusive **OR, NAND, AND, NOR,** and **OR,** functions.

Figure 7-11 shows the pin configuration for the 74S381 arithmetic logic unit/function generator. This **IC** is a Schottky TTL ALU that performs eight binary arithmetic/logic operations on two 4-bit words, as shown in Table 7-5. The operations shown in the function table (see Table 7-5) are selected by the three function-select lines (S_0, S_1, S_2). *Note:* If you place a low logic level on all three inputs (pins 5, 6, and 7, Figure 7-11) it will force all F outputs (pins 8, 9, 11, and 12) to be lows.

SELECTION			ARITHMETIC/LOGIC OPERATION
S_2	S_1	S_0	
L	L	L	CLEAR †
L	L	H	B MINUS A
L	H	L	A MINUS B
L	H	H	A PLUS B
H	L	L	A ⊕ B
H	L	H	A + B
H	H	L	AB
H	H	H	PRESET††

†FORCE ALL F OUTPUTS TO BE LOWS
††FORCE ALL F OUTPUTS TO BE HIGHS

Table 7-5: Function for the 74S381 ALU (Courtesy Monolithic Memories)

The standard test load for this **IC** is shown in Figure 10-7. However, in this case, R_L = 280 ohms, C = 15 pF, and you can eliminate switch S_1. Propagation delay time from input to output is typically 10 nanoseconds. Table 7-6 contains the electrical characteristics over the listed operating conditions.

Binary Arithmetic in a Computer System

You have seen how binary numbers are subtracted on paper and how they are subtracted by adding the 2's complement of the subtrahend to the minuend and ignoring the carry bit. You will

ELECTRICAL CHARACTERISTICS OVER OPERATING CONDITIONS

PARAMETER	TEST CONDITIONS		MIN	TYP	MAX	UNIT
LOW-LEVEL INPUT VOLTAGE					0.8	V
HIGH-LEVEL INPUT VOLTAGE			2			V
INPUT CLAMP VOLTAGE	V_{CC} =MIN 1_1=−18mA				−1.2	V
LOW-LEVEL INPUT CURRENT	V_{CC} =MAX V_1=0.5V	ANY S INPUT			−2	mA
		Cn			−8	
		ALL OTHERS			−6	
HIGH-LVEL INPUT CURRENT	V_{CC} =MAX V_1=2.7V	ANY S INPUT			50	μA
		Cn			250	
		ALL OTHERS			200	
MAXIMUM INPUT CURRENT	V_{CC} =MAX V_1=5.5V				1	
LOW-LEVEL OUTPUT VOLTAGE	V_{CC}=MIN V_{1H}=2V V_{1L}=0.8V 1_{0L}=20mA				0.5	V
HIGH-LEVEL OUTPUT VOLTAGE	V_{CC}=MIN V_{1H}=2V V_{1L}=0.8V 1_{0H}=−1mA		2.7	3.4		V
OUTPUT SHORT-CIRCUIT CURRENT*	V_{CC} =MAX		−40		−100	mA
SUPPLY CURRENT	V_{CC} =MAX			105	160	mA

*NOT MORE THAN ONE OUTPUT SHOULD BE STORED AT A TIME.

Table 7-6:   Electrical characteristics of the 74S381 over various operating conditions (Courtesy Monolithic Memories)

remember that using 2's complement simplifies IC circuits. Virtually all microprocessors have ALU's (you may find that some other designation is used but, generally, an ALU is included) and, as you now know, the ALU performs arithmetic and logic operations. This is done on the data bytes in a computer.

Using only the basic adder (which the ALU has), a computer programmer can write routines that will subtract, mulitply, and divide, giving the microprocessor system complete arithmetic capabilities. However, as you know, most ALU's provide other built-in functions, including Boolean algebra, logic operations, and shift capabilities.

To review binary basics and see how an 8-bit computer system performs the various arithmetic operations, study the following examples.

Binary Subtraction:

In a computer, subtraction is performed exactly as has been explained in previous pages of this chapter. For example:

an 8-bit computer	decimal
00001010	10
+ 11111101 (2's complement)	− 3
00000111	7

The ninth bit (final carry bit) that results from adding the two binary numbers is not required since this bit is not carried. This is easily accomplished in an 8-bit computer system because the circuits simply are not capable of carrying more than 8 bits.

Binary Multiplication:

In a computer system, this function is performed using a form of addition shifting. Therefore, all we need are adder circuits and shift registers. There are only three rules for binary multiplicaton. These are:

$$0 \times 0 = 0 \qquad 0 \times 1 = 0 \qquad 1 \times 1 = 0$$

As an example of their use, assume that we want to multiply decimal 2 × 21. This would be summed as follows:

```
      10101
  ×      10
      00000
      10101
     101010
```

Now, by examining this example, you can see that this is a form of addition and shifting. Next, referring to the binary conversion table given in Table 2-1, you can see that we use shift operations to multiply binary numbers by powers of 2, $2^0 = 1$, $2^1 = 2$, $2^2 = 4$, and so on. A shift to the left of the 1-bit position will multiply by 2 (refer to Table 2-1). A shift to the left of two positions mulitplies by 4, etc. Of course, a right shift of one position divides by 2, a right shift shift of two positions divides by 4, and so on.

Binary Division:

The division rules for binary are even simpler than the multiplication rules. There are only two. These are:

$$0 \div 1 = 0 \qquad\qquad 1 \div 1 = 1$$

As explained in the section on division, a computer simply reverses the multiplication process.

CHAPTER 8

How to Select and Use Memory IC's

Real-world, hands-on experience with practical examples — that's how this chapter helps you save time and produce good results when working with random access memories (**RAM's**) and read-only memories (**ROM's**). You will learn how **RAM's** are used to store data for use by electronic systems such as microcomputers.

You will also understand the semiconductor memories used to permanently store computer instructions. As explained in Chapter 1, this **IC**, a **ROM**, is a memory from which data can be read out repeatedly. This chapter, with its low-cost, effective methods and techniques, is a ready-to-use Reference Guide you can refer to daily.

How and Why Memory IC's Are Used

As seen in Chapter 1, a memory circuit in digital electronics is a circuit that can store a 1 or 0 bit. The storage units (flip-flops) in a memory **IC** are generally referred to as *memory cells.* Microprocessors use binary data (words) to communicate with other elements in a system. As a general rule, microprocessors are used with memories (**RAM's** and **ROM's**, typically both), that hold binary data to be manipulated by the microprocessor, and instructions to be followed by the microprocessor during the program. You will find that the data is usually held in **RAM IC's**, and instructions are generally stored in **ROM's**.

When selecting a **RAM** for a certain use, you can choose either a *static* or *dynamic* type. A typical dynamic **RAM** stores information, as an electrical charge, on the gate capacitance of a metal oxide semiconductor (**MOS**) transistor. Now, as you know, all transistors have some leakage current, which means the device will eventually lose the stored information unless some means is provided to periodically bring the charge up to its proper level. This is exactly the same type problem that is encountered when using a static shift register (see Figure 1-23).

There are several ways to improve the information holding time of a dynamic **RAM**. However, the circuitry is complex and you usually end up building circuits external to the **IC**, so why would you use a dynamic **RAM**? If your requirements are for low power consumption and high speed operation, the dynamic **RAM** is best. On the other hand, if your requirements are not so demanding and you can get by with slower operation and more power consumption, your best bet is to use a static **RAM**. These **IC's** store the binary information on flip-flops, or a latch, and have good information holding time, without the need for external information refreshing circuits.

By the way, many **RAM's** are volatile. In other words, you lose all information stored in the **IC** when power is removed. The safest way to go, in this case, is to use a battery-maintained power supply when the equipment is placed in a standby condition. If you do this, your best choice of a memory **IC** is a dynamic **RAM** because, as mentioned, they require less power than the static **RAM's**.

Understanding and Using

Random Access Memories

The 7489 **TTL IC** is a 16 × 4 **RAM**. In order to understand what is meant by this statement, we must examine *serial/parallel expansion* of memory cells, i.e., flip-flops. Memory **IC's** are built around a system of basic 1-bit addressable flip-flops that can be set to a 1 or 0 state, permitted to hold either state for a specified length of time and then provide that stored level on demand. For our purpose, the basic memory cell will be represented by a D type flip-flop. Chapter 2 describes the characteristics of D type flip-flops. See

Figure 2-17 for a basic logic diagram and Figure 2-18 for an operational table and example timing diagram.

Our example flip-flop is only one of many identical flip-flops in a memory system. Therefore, a single flip-flop must include a method for accessing one particular flip-flop of the group. As illustrated in Figure 8-1, the operation of the cell can be selected and controlled by means of various inputs to a couple of **AND** gates.

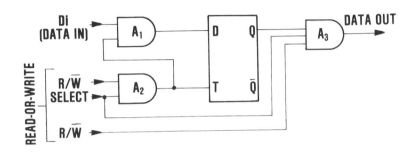

INPUTS			MODE
SELECT	R/W̄	Dₗ	
0	X	X	HOLD, S=0
1	0	0	0 INTO MEMORY S = 0
1	0	1	1 INTO MEMORY S = 1
0	1	X	READ, S = STORED Dₗ BIT

Figure 8-1: D type flip-flop memory cell and function table

As you can see, this basic cell has several inputs and one output. The inputs are a data terminal, cell select, and two read or write terminals. *Note:* Since, in actual use, there are many of these cells, there must be some method of selecting a particular cell. In our example, this cell is selected by setting the select input to a logic high (1), as can be seen by referring to the function table shown in Figure 8-1. Actually, the operation of the memory cell (select and control) is brought about by the inputs applied to the **AND** logic gates, A1, A2, and A3. See the logic diagram, Figure 8-1.

We know that this particular cell has four inputs: data in (D_j), select, and two read/write. Note, the read-or-write input labeled $\overline{R/W}$ is the complement of the input $R/\overline{W}$, and W is the complement of R, as shown by $R/\overline{W}$.

By referring to Figure 1-16, which shows a logic diagram for a 7488 read-only memory, and Figure 8-1, you can see that, typically, both **RAM**'s and **ROM**'s are controlled by selecting certain input logic levels to four or five gates. Now, let's refer back to the function table shown in Figure 8-1 and see how the memory cell shown in this figure works.

First, to select this particular cell you must place a logic high on the select input. See the left-hand column of the truth table. Next, note that all inputs and the output are blocked when you place the select input at a logic low (0). This is the *hold* mode where all information is stored and protected from any change due to other incoming data. To open the cell for read-or-write functions, you need to place a logic high on the select input, as we have explained. But, you must choose whether you want one or the other. To write a logic high or logic low into memory, the $R/\overline{W}$ input must be placed at a logic low (and $\overline{R/W}$ should be logic high). When you set the $R/\overline{W}$ input to a logic low, you cut off **AND** gate A3, which results in a logic low output. On the other hand, placing the complemented logic high level at $\overline{R/W}$ enables gates A2 and, in the end, gate A1.

At this point, you have both inputs to gate A2 at a logic high. In other words, you have both inputs to the flip-flop (D and T) enabled. Now, whatever data is presented to the flip-flop T input will be presented to the flip-flop Q output. In this mode of operation, the Q output will be at a logic low if the data input is at a logic low, or Q = 1 if data input = 1. The second and third lines of the function table in Figure 8-1 contain this same information.

Your next step after the data input operation is completed, is to set the select input to a logic low. When you do this, it causes the stored data (logic low or high) to remain at the Q output of the flip-flop. However, it will not appear at **AND** gate A3 output until you enable the cell by setting the read and write to input to select = 1, $R/\overline{W}$ = 1. This circuit (see Figure 8-1) would be called a 1 × 1 **RAM**. Of course it is a very simple, basic cell, but it gives you the idea of how the more complex memory systems operate.

You will remember that in the beginning of this section about **RAM**'s, it was stated that one must understand serial/parallel expansion of memory cells before a term like 16 × 4 **RAM** would have much meaning. As with all other serially operated systems, basically, a serially wired memory has only a single data input and one output; that is, it can accept or read out only one bit at a time.

It is possible to assemble any number of flip-flop memory cells in such a way that one of them can be used in a reading or writing operation. For example, with two data inputs and one output, we have what is called a *serial 2 × 1 memory format*. This setup requires two flip-flop memory cells. Using four flip-flops and associated circuits with one output, may result in a 4 × 1 format. Remember, we are still only able to handle one bit at a time on the output.

Next, let's consider using parallel hookups for expansion of memory. This system permits us to work with more than one bit at a

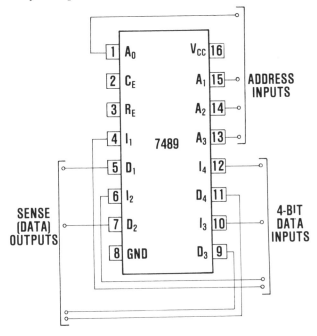

$\overline{CE}$	R/E	OUTPUTS	MODE
0	0	ALL 1's	WRITE
0	1	COMPLEMENT OF STORED D's	READ
1	X	ALL 1's	HOLD

Figure 8-2: 7489, 64-bit random access read/write memory pinout and function table. This device is a 16 × 4 TTL RAM

time. Even better yet, using both serial and parallel during construction of a memory system will make it possible to handle 4-bit words. The 7489, 64-bit **RAM** discussed in Chapter 1 is an **IC** manufactured using both serial and parallel circuits (see Figure 1-17). The 7489 is organized as a 16 × 4-bit word **RAM** capable of storing up to 16 different 4-bit words. It has 64 separate memory cells arranged in a 16 x 4 parallel/serial manner that can read out 16 different 4-bit words. Figure 8-2 shows the pinout and function table for this **IC**.

Notice that there are four data inputs (I_1 through I_4), and four data outputs (D_1 through D_4). Now, remembering that this **IC** has 16 word storage locations, it is only reasonable to expect it to have four binary address lines. These are labeled A_0 through A_3. Incidentally, some manufacturers use different letters to indicate inputs, outputs, and the like. For example, you will find the data outputs (also called *sense* outputs) labeled S_1 through S_4 in many technical articles, books, etc. However, the pin numbers for the **IC**'s will always be the same, regardless of who manufactured the **IC**.

There is an R_E (sometimes labeled R/$\overline{W}$) control terminal that you use for the purpose of writing into memory by use of the data inputs, or reading out via the data outputs. For instance, the truth table (see Figure 8-2) shows that if you place R_E at a logic level 0, data is written into memory. Or, the complement of any binary number you have previously entered is read out if you place R_E at a logic 1. The C_E (sometimes labeled M_E) terminal is the memory enable control input. The truth table shows us that you should set the C_E terminal to a logic low for either a reading or writing operation. Or, if you wish to place the memory in the holding mode, place a logic high on the C_E terminal. In this case, all outputs are set to a logic high. In summary, to *place* a 4-bit word into memory:

Step 1. Set C_E to logic 1.

Step 2. Set the address location at A inputs.

Step 3. Set the data to be stored at I inputs.

Step 4. Set R_E to logic 0.

Step 5. Set C_E to logic 1 (to enable the **IC**).

When you perform Step 5, the data you placed at the I inputs will be written into the address you specified at the address inputs (Step 2).

To *read* a 4-bit word out of the **IC**:

Step 1. Address the same location as you entered in Step 2 in the enter data procedure, via the A inputs.

Step 2. To enable the **IC**, set R_E to logic 1.

Step 3. At this point, you should be able to find an inverted 4-bit word on the D outputs; i.e. in respect to the 4-bit word you previously stored in memory, a complemented 4-bit word.

To perform either read or write (as previously explained) with this **IC**, requires pull-up resistors on each output terminal (D_1 through D_4). These resistors should be connected between the ouput pins and V_{cc}. See Experiment 8-3.

The access time and total storage capacity are the two most important characteristics of **RAM** devices. The access times are related to the clock frequency and memory capacity — the higher the capacity, the longer the access time. Significant steps in **IC** memory

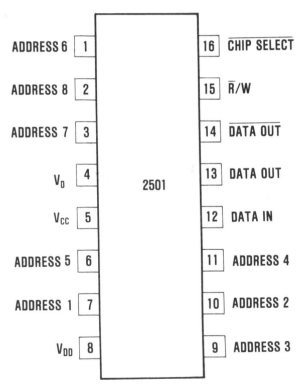

Figure 8-3: Package and pin configuration for a 256 × 1 static read/write RAM (2501). $V_{cc} = +5$, $V_{DD} = V_D = 9V$

capacity have been made; for example, 256 to 1024 and from 1024 to 4096-bit, and so on, per **IC**. Furthermore, prices are comparatively low. A 4027-P2, 4k × 1 dynamic **RAM** with an access time of 150 nanoseconds, sells for a few dollars. For experimental purposes, you may want a less expensive **RAM**. In general, you can find a 256 × 1 static read/write **RAM** that is quite inexpensive. Figure 8-3 shows the package and pin configuration for the relatively low-cost 2501.

Maximum power dissipation is listed as 1.6 mW/bit; however, this is required during read or write only. For standby operation, 150 μW/bit is obtained by removing V_D and reducing V_{DD} to -4.0 V. Removal of V_D alone will cut power dissipation by a factor of 1.5. Table 8-1 is a list of maximum ratings. *Stresses above those listed may cause permanent damage to the IC.*

MAXIMUM RATINGS	
OPERATING TEMPERATURE	0°C to +70°C
STORAGE TEMPERATURE	−65°C to +150°C
ALL INPUT OR OUTPUT VOLTAGES WITH RESPECT TO THE MOST POSITIVE SUPPLY VOLTAGE, VCC (+5.25V)	0.3V to −20V
SUPPLY VOLTAGES V_{DD} and V_D WITH RESPECT TO V_{CC}	−18V
POWER DISSIPATION AT TA = 70°C	640mW

Table 8-1:   Ratings table for a 256 × 1 static read/write RAM. All voltages referenced to ground

All inputs of this type **RAM** can be driven directly by standard bipolar integrated circuits (**TTL, DTL,** etc.). The data output terminals (from internal buffers) are capable of sinking 1.6 mA, sufficient to drive one standard **TTL** load; for example, a **NAND** gate such as the 7400. To do this, connect an **AND** gate to each output (pins 12 and 13), as shown in Figure 8-4.

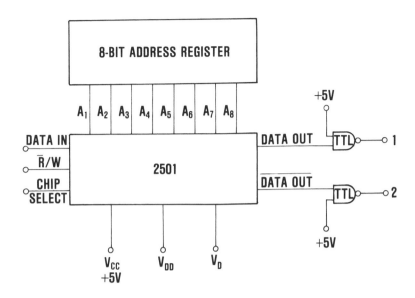

Figure 8-4:   Test setup for checking a 256 × 1 static read/write RAM

How a Read-Only Storage IC
Is Used and Programmed

Although a microprocessor **IC** performs many of the functions such as arithmetic and control of a computer, it usually does
not contain the memories and input/output (I/0) functions of a computer. There are **IC's** — called *computer on a chip* — that contain all
these basic functions. But for our example, we shall restrict our explanations to single-chip microprocessors.

In most systems, at least the ones we are referring to, a **ROM**
is required to store the microprocessor program or instructions. As a
matter of interest, you will find microprocessor-based systems in
computers, industrial control systems, video games and in the
dashboard of many cars. For example, the 1981 Chrysler Imperial
uses two microprocessors.

Of all the **ROM** families, perhaps the programmable read-
only memory **(PROM)** is the most useful to the experimenter. You
will recall that the fusing operation calls for addressing the location
and then running a few milliamperes of current through a special
programming connection (see Figure 1-18). When you perform this

operation, the outputs at each address are set permanently to your selected pattern of 1's and 0's. As an example of the programming procedure of a so-called *field programmable* **ROM,** we will use a Signetics 8223. This **IC** is a **TTL** 256-bit **ROM** organized as 32 words with 8 bits per word, and is the same as several other **PROM'**s distributed by the industry.

EXPERIMENT 8-1

PROM Programmable Procedure

Using an 8223

The workbench setup for programming the 8223 is again shown in Figure 8-5 so it will be convenient and easy to follow the programming procedures.

Programming Procedure

Comments: When laying out the test circuit shown in Figure 8-5, keep your lead lengths as short as possible, particularly the ones going to your power supply. Also a capacitor of 10 microfarads *minimum* should be connected from the +12.5V line to ground. Place this capacitor as close as practical to the **IC** being programmed—off pin 16 is good (see Figure 8-5). The 8223 standard part normally is shipped with all outputs at a logic 0. To write a logic 1, perform the following eleven steps.

Step 1. Read through the entire procedure before starting the actual programming.

Step 2. Start with pin 8 (see Figure 8-5) grounded and V_{cc} removed from pin 16.

Step 3. There should be no loads connected to any output. If there are, remove them.

Step 4. Ground chip enable, pin 15.

Step 5. Address the desired location by applying ground (0 to 0.4V maximum) for a 0 and +5.0V (+2.8 minimum) for a 1 at the address inputs (the five binary address lines are pins 10 through 14).

Step 6. Apply +12.5V ± 0.5V to the output to be programmed through the 390 ohm ± 10% resistor by setting the switch in the program position. Note: program one output at a time.

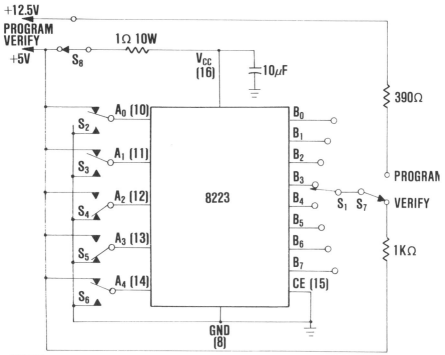

NOTES:
1. THE 10μF CAPACITOR ACROSS PIN 16 TO GROUND IS REQUIRED TO ELIMINATE NOISE FROM VCC.
2. DURING PROGRAMMING SWITCH S_8 MUST BE IN THE VERIFY POSITION LONG ENOUGH FOR THE 10μF CAPACITOR TO DISCHARGE TO 5.0 VOLTS.

Figure 8-5: Manual programmer diagram for the 8223 field-programmable ROM. Note: This IC is the same as Intersil 5600, M.M. 6330, T.I. SN7488

Step 7. Wait for a short period and then apply + 12.5V to V_{CC} (pin 16) and *remove as quickly as possible.* You want to limit the overshoot to no more than 1.0V. Use a simple diode clamping circuit, if necessary. The idea is to keep the rise time as short as possible — 50 μsec or less.

Step 8. Verify that you have programmed a bit by applying 5 volts to V_{CC} and 5 volts through a 1 k ohm resistor to the output. See program verify switches shown in Figure 8-5.

Step 9. Proceed to the next output and repeat the steps, or change address and repeat.

Step 10. Continue until you have the entire bit pattern programmed into the **PROM.**

Step 11. After you have the **IC** programmed, check to be sure all bits are entered (the code is correct). If you find that a bit has not been programmed, return to that bit and repeat the programming procedure once.

EXPERIMENT 8-2
Programming a PROM with a Truth Table

Techniques for programming **PROM**'s differ according to the technologies used to implement the device. Certain types of **MOS PROM**'s (**EPROM**'s described in Chapter 1) can be erased and reprogrammed, but bipolar **TTL PROM**'s can be programmed only once, as has been explained. Now, Signetics' 8223 normally comes with all outputs set at a logic 0, but with some **PROM**'s, burning the internal fuse open will represent a stored 0, whereas in others (such as the 8223), it will represent a stored 1. Because of this, it is very important that you have program instructions, specification sheets, etc., for the **PROM** you are working with.

If you are programming a **PROM** from scratch, a truth table must first be drawn up. You will also use this truth table when the stored information is being read. The address code depends on the equipment in which the **PROM** is used (a computer or whatever). As an example, a partial truth table that might be used to program the **PROM** shown in Figure 8-6 is shown in Table 8-2.

PROGRAM SEQUENCE	A_8	A_7	A_6	A_5	A_4	A_3	A_2	A_1	A_0	O_8	O_7	O_6	O_5	O_4	O_3	O_2	O_1	O_0
			ADDRESS INPUTS										OUTPUTS					
1	0	0	0	0	0	0	0	0	0	0	0	0	0	0	0	1	1	
2	0	0	0	0	0	0	0	0	1	0	0	0	0	0	1	0	1	
3	0	0	0	0	0	0	0	1	0	0	0	0	0	0	1	1	1	
4	0	0	0	0	0		0	1	1	0	0	0	0	1	0	1	0	
•	•	•	•	•	•	•	•	•	•	•	•	•	•	•	•	•	•	•
•	•	•	•	•	•	•	•	•	•	•	•	•	•	•	•	•	•	•
•	•	•	•	•	•	•	•	•	•	•	•	•	•	•	•	•	•	•

Table 8-2:   A partial truth table that might be used to program a PROM with the setup shown in Figure 8-6

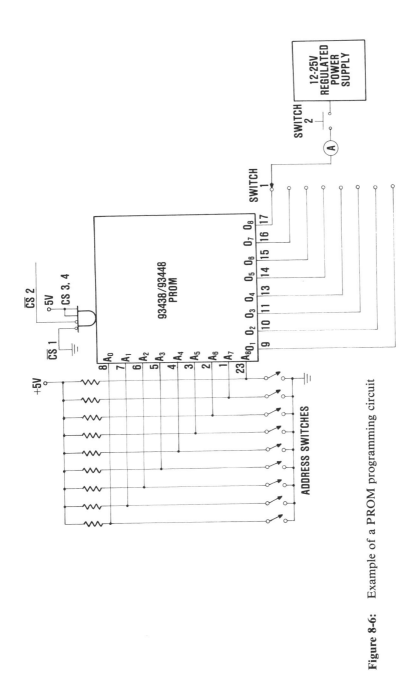

Figure 8-6: Example of a PROM programming circuit

Again, it must be emphasized that this is only an example of how the programming of a **PROM** is accomplished and you probably will have to refer to the manufacturer's data book, spec sheet, or some other such instructions for the specific **PROM IC** you want to program. A program setup is illustrated in Figure 8-6. As before, logic switches are used to select the proper address. In general, these switches do not need to be debounced. You must also apply a programming pulse to the output pin associated with the bit to be programmed. More than likely, you'll find that the other outputs may be left open or tied to any high. *Note:* In almost every case, when using manually operated switches to program one bit at a time, you should not try to program more than one output at a time.

The procedure for actual programming, using the setup shown in Figure 8-6, would be:

Step 1. Apply V_{cc} and ground to the **IC**.

Step 2. Use the address switches to select the desired address. The example truth table shows all switches are to be set at a logic 0.

Step 3. Set SWI to output 0.

Step 4. Depress SW2 for a very short time (less than ½ second, if possible) and release. Watch the current meter to see that you do not exceed the manufacturer's recommended value (see Table 8-3 for an example). This step is to provide the amount of current necessary to permanently open a nichrome link (other materials are used for these fuse links, however the nichrome fuse link is the most popular).

Step 5. When you burn the fuse link open to produce a logic 0 or logic 1, depending on the circuit design, a considerable amount of on-chip heat is developed. Therefore, in all cases, allow several seconds for the chip to cool down before you bring the readout pin to the proper level for a verify check.

Step 6. Set SW1 to output 02 and repeat Steps 4 and 5.

Step 7. Now, referring to Table 8-2 (program sequence 1), we see that the word 11000000 is now programmed in address A_0 through A_8. You should continue to select each address and program each word (one bit at a time) by burning open each of the fuses according to your truth table, until you have completed programming the **PROM IC**.

FPLA's and PAL's

Field programmable logic arrays (**FPLA**) and programmable array logic (**PAL**®, a trademark of Monolithic Memories, Inc.) chips have one thing in common with field programmable **ROM's**. They all can be programmed with a **PROM** programmer. The **FPLA** and **PAL** chips contain arrays of logic gates interconnected via the same kind of fusable links used in **PROM's**.

Using the same basic idea as shown in Figure 8-6, fusable links can be opened in various patterns to produce an IC of your design. The idea, as you'll remember, is to provide enough current to permanently open the link, thereby programming a bit to a logic 0 or 1, depending on the circuit design.

Carrying this idea on further, we can illustrate the programming of a **PROM** of four 2-bit words by using a logic diagram. Figure 8-7 shows a stripped-down version of the **PROM** logic diagram containing an **AND** gate array followed by an **OR** gate array.

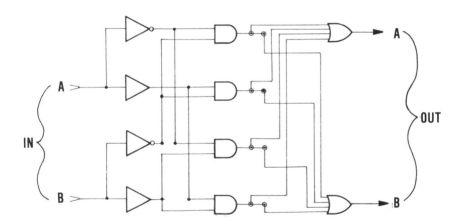

Figure 8-7:   Logic diagram of a PROM using four 2-bit words. A solid dot at an array line means the connection is fixed, i.e., programmed when the chip was manufactured. The small circles with a dot in the center indicate a fusable link

As is readily apparent from Figure 8-7, the **AND** array is permanently programmed while the **OR** array is left for the user to program. The programming is done according to the standard 00, 01, 10, 11 input sequence already explained. Now, the **PAL** (Figure 8-8)

works similar to a **PROM**. The **OR** array is factory-programmed. The result is that many of the most commonly used logic functions are available, using your design.

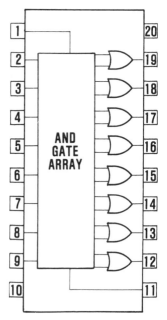

Figure 8-8:   Pin configuration and logic diagram for a PAL10HB (Courtesy Monolithic Memories, Inc., 1165 E. Arques, Sunnyvale CA. 94086)

The next step up the line, in both customized design and *price,* is the **FPLA**. In this chip, both the **AND** and **OR** arrays are programmable. Although the **FPLA**, like the **PAL** and **PROM**, can be programmed using the PROM programmer, it is about twice as much work due to the fact that both arrays need to be programmed.

Programming Erasable PROM's

The erasable **PROM (EPROM)** is another chip that the electronics experimenter, hobbyist, and home computer owner should become familiar with because these very useful devices are being increasingly accepted as a circuit element. For example, Ultra-Violet Products, Inc., San Gabriel, CA 91778 is selling an **EPROM** erasing lamp that is low-cost and designed specifically for the small-system user, computer hobbyist, and experimenter. This high-powered unit will erase up to four chips at one time in as little as 20 minutes and it's easy to use. Just plug into the nearest outlet, load with chips and go! The entire unit weighs only one pound. *Warning:* Short-wave ultraviolet light can cause damage (the same as electric arc welding or sunburning) to your eyes and skin. Do not look directly into an **EPROM** erasing lamp. Also, avoid shining an ultraviolet lamp on reflective surfaces. Special safety goggles are available from such companies as Ultra-Violet Products, Inc.

Just as there are closed or fused-open junctions in a **PROM** array, the **EPROM** uses static charges on **MOSFET** transistors to achieve the effects of an open or closed junction. As an example, one type of **EPROM** made by National Semiconductor is the 2048-bit MM5203, packaged in a 16-pin DIP with a quartz window on top. The quartz window is transparent to short-wave ultraviolet light (2530 Angstrom units). To erase the unwanted data, you simply expose the window to an ultraviolet device. For proper erasure, the chip within the **IC** package must be exposed to strong ultraviolet radiation for a few minutes (typically, about 15 to 20 minutes).

As another example, to erase one of Motorola's 1024 x 8 **EPROM**'s (they call them *alterable-ultraviolet-*ROM's), you would expose the transparent lid shown in Figure 8-9 to a dose of at least 15 watts per second per square centimeter of ultraviolet light, for 15 to 20 minutes. The rating of the ultraviolet lamp should be 12000 μW/CM² and the **IC** is placed within one inch of the lamp.

ULTRA VIOLET LIGHT WINDOW

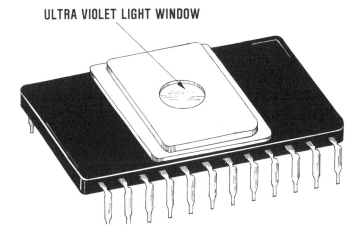

Figure 8-9:   1024 x 8 alterable (ultraviolet) ROM

Inside the **IC** is a special type transistor called a *floating gate avalanche-injection metal oxide semiconductor,* which is abbreviated **FAMOS.** The gate has no external connection. When you are *programming* this device, a voltage in excess of -30 volts is usually required and is placed across the drain and source. This results in an injection of high-energy electrons into the floating gate. The negative charge on the gate allows current to flow between source and drain during read-out. In addition to the usual address inputs, data outputs, decoders and buffers, an **EPROM** usually has a "program line" input terminal. An appropriate signal on the program terminal permits information to be programmed into the address you select.

EAROM and RMM

The electrically alterable read-only memory **(E/ROM)** and read-mostly memory **(RMM)** are used much as a **ROM** or **PROM** except that the data contained within the memory may be erased using electrical currents instead of ultraviolet light. The programming is similar to that of the **EPROM** but the erasure is a slow process, usually requiring external circuits which in general, are so slow that they are not practical for read/write use. For these reasons, you will find **PROM**'s and **EPROM**'s etc., are more frequently used in today's digital equipment.

EXPERIMENT 8-3

Using and Testing a Read/Write Memory IC

Figure 8-10 shows a simple circuit you can build to test a 7489 64-bit read/write memory **IC.** As with most digital **IC** test circuits, as long as you leave the data and address switches open, you are placing a logic 1 on their respective input terminals. Or, when you close one of the switches, you are placing a logic 0 on the respective input terminal.

The memory-read and write toggle switch is used to place the **IC** into either the read or write mode of operation. When you leave this switch open, you are placing a logic high on pin 3 (placing the **IC** in the read mode). Closing the switch places a logic low on pin 3 (changes the operation to write mode). The push-button switch (memory enable) must be depressed before either the read or write operation can be performed, but *not while changing the address.* If

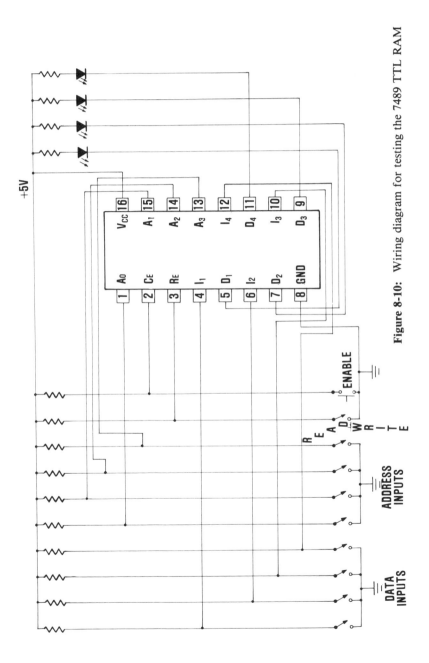

Figure 8-10: Wiring diagram for testing the 7489 TTL RAM

you do not follow these directions, you are sure to end up with some confusing data on the outputs that has no resemblance to your truth table outputs. Table 8-3 is a truth table for a 7489 **RAM.**

OPER-ATION	ADDRESS LOCATION	$\overline{ME}$ PIN 2	$R/\overline{W}$ PIN 3	OUTPUTS $D_2 - D_4$
READ	$A_0 - A_3$	0	1	0's FOR STORED DATA
WRITE	$A_0 - A_3$	0	0	ALL 1's

Table 8-3: Truth table for a 7489 RAM

A Simple Test:

Step 1. Apply V_{CC} to pin 16, ground to pin 8. Remember that this **IC** is a volatile memory. Do not remove power during this or the next test procedure.

Step 2. Write a logic 0 at all address switches.

Step 3. Read all locations. You should read a logic low at all of them. If you don't, the **IC** is probably defective.

Step 4. Write a logic 1 at all address switches. You should read a logic high at all locations during this part of the test. If not, the **IC** is probably defective.

Testing All 16 Locations:

Step 1. Make up a truth table for the binary data you wish to store at each of the 16 locations.

Step 2. Select the desired address by setting the address switches to the proper position.

Step 3. Be sure you place the switch on pin 2 (enable) to a logic low.

Step 4. Switch the read/write switch (on pin 3) to a logic 1 and then back to logic 0. The LED's should blink (turn off and on) indicating the binary data you entered is stored at the selected address.

Step 5. Repeat the preceding steps for the other 15 address locations. *Note:* Remember, the 7489 output is the *complement.* To increase the word length, you can use two or more 7489's. As an example, Figure 8-11 shows the basic hookup for using two of these **IC's**.

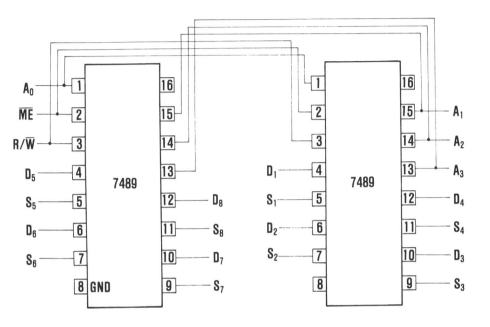

Figure 8-11: Wiring diagram for using two 7489 IC's to increase word length (32 x 4)

The address and read/write operations require a few more steps than when using a single **IC**. These are:

Step 1. To read the first 16 locations, address A0 to A3, place $\overline{ME}$ - 1 high, $\overline{ME}$ - 2 low, to enable **IC1**. You should see the data stored on pins 5, 7, 9 and 11 of **IC1**, or at the sense outputs.

Step 2. To read the second 16 locations, address inputs A4 through A7, place $\overline{ME}$ - 1 high and $\overline{ME}$ - 2 low, to enable **IC2**. You should see the outputs at pins 5, 7, 9 and 11 of the second **IC**, or at the sense outputs.

Step 3. The write operation: Select the proper pin 2 input, address, and the binary data you wish to be stored. Place this data at inputs D_1 through D_4 when you set pin 3 to a logic low. Although the wiring diagram shows the address pins of each **IC** as separate inputs, they actually can be tied together because only one **IC** is operational at a time, i.e., when its $\overline{ME}$ input is placed on a logic low. *Caution:* Be sure the **IC** you are not loading has its $\overline{ME}$ input at a logic high (open).

One thing you should watch when you select **IC2**, the memory location 17 is addressed 0000 and, when you program the following locations, they are addressed respectively up to locations 32. This locations' address should be 1111.

CHAPTER 9

Using Decoders/Drivers, Priority Encoders and LED Displays In Digital Systems

In this chapter, you will find ready-to-use tips and techniques that you can use to solve, or to avoid, problems when working with decoders and encoders, time-saving tips for digital decoder/display driver systems, and actual descriptions with pin connections for numerous LED displays.

You will learn which type driver is required for common anode LED displays and which one you'll need for common cathode units. Not only will you learn which type, but you will be able to connect a LED display into a circuit using actual pin layouts. For instance, Table 9-5 contains descriptions and pin configurations for Texas Instruments, Dialight, Litronix, and Monsanto LED displays.

What You Should Know About
Binary-to-Decimal Decoders

The basic circuit used to make a decoder is called a *matrix*. For our purpose, a matrix is an array of logic gate connections constructed inside an **IC** connecting integrated inverters and **AND** or **NAND** gates. The logic diagram of a BCD-to-decimal decoder (7442) is shown in Chapter 2, Figure 2-3. However, at this time, we are interested in a truth table as well as the logic diagram. These are both shown in Figure 9-1.

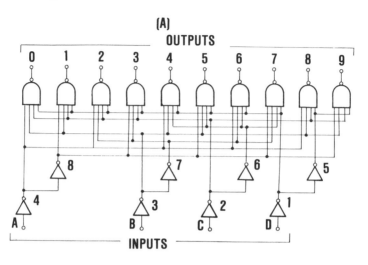

(A)

(B)

| BCD INPUT | | | | DECIMAL OUTPUT | | | | | | | | | |
D	C	B	A	0	1	2	3	4	5	6	7	8	9
0	0	0	0	0	1	1	1	1	1	1	1	1	1
0	0	0	1	1	0	1	1	1	1	1	1	1	1
0	0	1	0	1	1	0	1	1	1	1	1	1	1
0	0	1	1	1	1	1	0	1	1	1	1	1	1
0	1	0	0	1	1	1	1	0	1	1	1	1	1
0	1	0	1	1	1	1	1	1	0	1	1	1	1
0	1	1	0	1	1	1	1	1	1	0	1	1	1
0	1	1	1	1	1	1	1	1	1	1	0	1	1
1	0	0	0	1	1	1	1	1	1	1	1	0	1
1	0	0	1	1	1	1	1	1	1	1	1	1	0
1	0	1	0	1	1	1	1	1	1	1	1	1	1
1	0	1	1	1	1	1	1	1	1	1	1	1	1
1	1	0	0	1	1	1	1	1	1	1	1	1	1
1	1	0	1	1	1	1	1	1	1	1	1	1	1
1	1	1	0	1	1	1	1	1	1	1	1	1	1
1	1	1	1	1	1	1	1	1	1	1	1	1	1

Figure 9-1: 7442 (A) logic diagram, (B) truth table

Referring to the input table shown in Figure 9-1, you will notice that the inputs are labeled D, C, B, and A, rather than A, B,

C, and D, as in some other **IC** truth tables. This notation is used for most **IC** decoder devices. The D is used to indicate the most significant bit (MSB) of the input and the A, the least significant bit (LSB). To give you an idea of what is taking place in a 7442 decoder, let's chase an input signal (binary word) through the device.

Starting off with DCBA = 0000, you'll see that all outputs from the inverters (1, 2, 3, and 4) are complemented, i.e., $\overline{D}$, $\overline{C}$, $\overline{B}$, and $\overline{A}$. Therefore, the outputs are all at logic 1. Next, the second set of inverters (5, 6, 7, and 8) receive the inverted binary word and invert it again, resulting in outputs of D, C, B, and A, which are fed to the matrix as logic 0's. What this all adds up to is that there are two words (1111 and 0000) being fed to the inputs of the **NAND** gates. Now, if you trace the connections of the **NAND** gate on the extreme left (output number 0), you will find that all the *inputs* to this gate are logic 1. Therefore, its output would switch to logic 0. Under these same conditions, you'll find that at least one other input on all other **NAND** gates is logic 0. For this reason, the truth table shows all other decoder outputs are at a logic 1 level.

The 7442 decodes a 4-bit BCD number to one of ten outputs. In the example we just followed through, it was shown that the decoder indicated that the input was the BCD code for decimal number 0 when the 0 output switch was set to logic 0. This same information is given in the truth table. Also, notice that the truth table shows that each input condition causes just one output to switch to logic 0. This is the output corresponding to the decimal equivalent number of the BCD code.

Question: What if you wanted to make up a decoder that sinks the decimal output to logic 1 instead of sourcing to logic 0? You can do this very easily. Simply replace the **NAND** gates shown in Figure 9-1 with **AND** gates. Your new truth table would now show logic 0's in place of each logic 1 shown in the truth table in Figure 9-1. All logic 0's would be replaced with logic 1's and that's all there is to it.

Another question: What if you wanted to drive a high current load with a BCD-to-decimal decoder? In this case, a 7445 or 74145 might be your best bet. These **IC**'s have identical pin layouts as the 7442 (see Figure 2-3) however, their outputs feature high current capabilities (80 mA). The minimum output breakdown voltage for the 7445 is 30 volts and, for the 74145 it's 15 volts. Incidentally, the 7442 outputs are buffered to sink 16 mA.

Two other BCD-to-decimal decoders that have been designed to provide the necessary high-voltage characteristics required for driving gas-filled cold-cathode indicator tubes, relays, or other high-voltage interface circuitry are the 7441 and 74141. Figure 9-2 shows the pin configuration and a truth table for the 7441. The voltage on any output is recommended to not exceed a maximum of 70 volts. Recommended supply voltage is 4.75V to 5.25V.

The 7441 and 74141 have the same pin configuration, as is shown in Figure 9-2. However, the 74141 is a BCD-to-decimal decoder/driver with blanking, and its truth table is quite a bit different. Table 9-1 is the truth table for the 74141. Full decoding is provided for all possible inputs stated. For binary inputs 10 through 15,

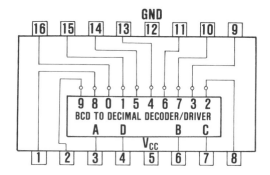

INPUT				OUTPUT
D	C	B	A	ON*
0	0	0	0	0
0	0	0	1	1
0	0	1	0	2
0	0	1	1	3
0	1	0	0	4
0	1	0	1	5
0	1	1	0	6
0	1	1	1	7
1	0	0	0	8
1	0	0	1	9

*ALL OTHER INPUTS ARE OFF

Figure 9-2: BCD-to-decimal decoder/driver 7441 pin configurations and truth table

all outputs are off. This **IC** was designed specifically to drive cold-cathode indicator tubes such as a NIXIE (Trademark of Burrough Corp). When the **IC** is connected to external digital circuits, the automatic blanking control ensures that the NIXIE tubes the **IC** is driving are not conducting for the time the next BCD coded word is shifted into the decoder. V_{cc} = 5V nom., power dissipation = 55 milliwatts (typically), output voltage is 65 (max.) and supply current V_{cc} (max) = 16 mA (max).

INPUT				OUTPUT ON*
D	C	B	A	
L	L	L	L	0
L	L	L	H	1
L	L	H	L	2
L	H	H	H	3
L	H	L	L	4
L	H	L	H	5
L	H	H	L	6
L	L	H	H	7
H	L	L	L	8
H	L	L	H	9
H	L	H	L	NONE
H	L	H	H	NONE
H	H	L	L	NONE
H	H	L	H	NONE
H	H	H	L	NONE
H	H	H	H	NONE

H = HIGH LEVEL, L = LOW LEVEL
*ALL OTHER OUTPUTS ARE OFF

Table 9-1:   Truth table for a 74141 BCD-to-decimal decoder/driver with blanking

Excess-3-to-BCD Decoder

The 7443 excess-3-code to decimal decoder is a **TTL MSI** array utilized in decoding and logic conversion applications. This **IC** decodes excess-3-code numbers to one of ten outputs. Figure 9-3 shows a logic diagram and pin configurations for the device.

The truth table of the 7443 decoder operation is shown in Table 9-2. Looking at this table, you'll find that the DCBA Input col-

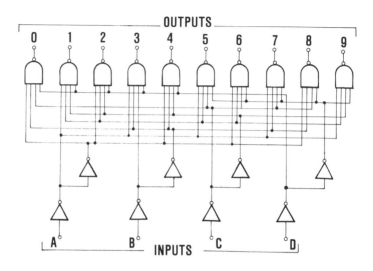

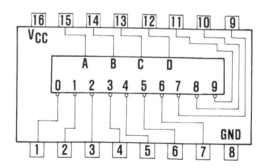

Figure 9-3:   Excess-3-to-decimal decoder. Logic diagram and pin configurations for the 7443

umns are labeled *excess inputs.* These input columns represent the excess-3-code (sometimes called XS3 *code*). There are many codes using the binary system, including the 2421, 5421, reflected gray, 2 out of 5, biquinary, and the XS3. Excess-3-binary coded decimal pertains to a code based on adding 3 to a decimal digit and then converting the result directly to binary form. The use of this code simplifies the execution of certain mathematical operations in a computer that must handle decimal numbers.

Looking at the other table (Decimal Output) shown in Table

EXCESS INPUT				DECIMAL INPUT									
D	C	B	A	0	1	2	3	4	5	6	7	8	9
0	0	1	0	0	1	1	1	1	1	1	1	1	1
0	1	1	0	1	0	1	1	1	1	1	1	1	1
0	1	1	1	1	1	0	1	1	1	1	1	1	1
0	1	0	0	1	1	1	0	1	1	1	1	1	1
0	1	0	1	1	1	1	1	0	1	1	1	1	1
1	1	0	0	1	1	1	1	1	0	1	1	1	1
1	1	0	1	1	1	1	1	1	1	0	1	1	1
1	1	1	1	1	1	1	1	1	1	1	0	1	1
1	1	1	0	1	1	1	1	1	1	1	1	0	1
1	0	1	0	1	1	1	1	1	1	1	1	1	0
1	0	1	1	1	1	1	1	1	1	1	1	1	1
1	0	0	1	1	1	1	1	1	1	1	1	1	1
1	0	0	0	1	1	1	1	1	1	1	1	1	1
0	0	0	0	1	1	1	1	1	1	1	1	1	1
0	0	0	1	1	1	1	1	1	1	1	1	1	1
0	0	1	1	1	1	1	1	1	1	1	1	1	1

Table 9-2: Truth table for an excess-3-to-decimal decoder (7443)

9-2, we find that at the outputs, the lead with the decimal number corresponds to the coded input switches to a logic 0. Also, notice that all other outputs stay at logic 1. Each output of the 7443 can sink a load that draws up to 16 mA.

Another **IC** decoder circuit with the pin layout exactly like the 7443 is the 7444 excess-3-gray decoder. This **IC** takes an XS3 gray code at the DCBA inputs and converts it to a decimal output. The logic circuitry inside the 7444 is like that of the 7442 and 7443 devices but the **NAND** gate input wiring is changed to correspond to an XS3-gray code. Table 9-3 is a truth table for the 7444.

Binary-to-Octal Decoders

Before we examine the binary-to-octal decoder **IC**, let's review a bit. To convert an octal number to its binary equivalent, you will

EXCESS 3 GRAY INPUT

D	C	B	A
0	0	1	0
0	1	1	0
0	1	1	1
0	1	0	1
0	1	0	0
1	1	0	0
1	1	0	1
1	1	1	1
1	1	1	0
1	0	1	0
1	0	1	1
1	0	0	1
1	0	0	0
0	0	0	0
0	0	0	1
0	0	1	1

DECIMAL OUTPUT

0	1	2	3	4	5	6	7	8	9
0	1	1	1	1	1	1	1	1	1
1	0	1	1	1	1	1	1	1	1
1	1	0	1	1	1	1	1	1	1
1	1	1	0	1	1	1	1	1	1
1	1	1	1	0	1	1	1	1	1
1	1	1	1	1	0	1	1	1	1
1	1	1	1	1	1	0	1	1	1
1	1	1	1	1	1	1	0	1	1
1	1	1	1	1	1	1	1	0	1
1	1	1	1	1	1	1	1	1	0
1	1	1	1	1	1	1	1	1	1
1	1	1	1	1	1	1	1	1	1
1	1	1	1	1	1	1	1	1	1
1	1	1	1	1	1	1	1	1	1
1	1	1	1	1	1	1	1	1	1
1	1	1	1	1	1	1	1	1	1

Table 9-3:   Truth table for a 7444 excess-3-gray-to-decimal decoder

remember that in Chapter 2 it was stated that you must convert each digit of the octal number to the binary equivalent by *using three binary digits per octal digit.* Another point: For three octal digits, it is necessary to use nine binary digits. On the other hand, to convert a binary number to its octal equivalent, you must divide the digits of the binary number into groups of three. Starting from the right . . . add 0 to the left to complete the digits, if necessary. For example:

added zero	010	100	111 (binary)
	2	4	6 (octal)

Now, from this brief review, it appears that a binary-to-octal decoder would require three lines of input. Figure 9-4 shows that this is true. Our example binary-to-octal decoder, the 8250 or MC7250P, converts three lines of input (A, B, and C) to one-of-eight output. The fourth input line (D) is utilized as an inhibit to allow use in larger decoding networks.

This **IC** accepts a BCD number at inputs A, B, C, D, and provides a low at one of eight outputs, while the other seven outputs remain high. For example, a 1010 input results in output 5 (pin 4) going low. Binary numbers from 1110 to 1111 at the input send all outputs

A

B

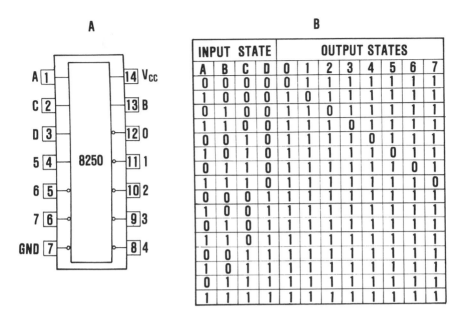

INPUT STATE				OUTPUT STATES							
A	B	C	D	0	1	2	3	4	5	6	7
0	0	0	0	0	1	1	1	1	1	1	1
1	0	0	0	1	0	1	1	1	1	1	1
0	1	0	0	1	1	0	1	1	1	1	1
1	1	0	0	1	1	1	0	1	1	1	1
0	0	1	0	1	1	1	1	0	1	1	1
1	0	1	0	1	1	1	1	1	0	1	1
0	1	1	0	1	1	1	1	1	1	0	1
1	1	1	0	1	1	1	1	1	1	1	0
0	0	0	1	1	1	1	1	1	1	1	1
1	0	0	1	1	1	1	1	1	1	1	1
0	1	0	1	1	1	1	1	1	1	1	1
1	1	0	1	1	1	1	1	1	1	1	1
0	0	1	1	1	1	1	1	1	1	1	1
1	0	1	1	1	1	1	1	1	1	1	1
0	1	1	1	1	1	1	1	1	1	1	1
1	1	1	1	1	1	1	1	1	1	1	1

Figure 9-4: Binary-to-octal decoder (8250). (A) is the pin configuration for a 14-lead dual in-line-molded. (B) is a truth table for the 8250 or MC7250P. The selected output is a logic 0

high. See the logic diagram for the 8250 shown in Figure 9-5 (page 232). Notice that the logic diagram for this **IC** is very similar to that of the 7442 BCD-to-decimal decoder (and all the rest of the decoders we have discussed). Almost all of these **TTL** decoders/drivers are much the same except for the internal wiring network, i.e., the matrix.

Priority Encoders

Encoders are usually considered as devices that convert other number systems into binary number systems. However, you will find the terms *encoder, decoder,* and *multiplexers* are often interchanged in technical literature. In a strict sense, a *decoder* converts from one code or numbering system to another, as we have seen. The term *encoder* is usually applied when an uncoded value is converted to a coded form, and the other term, *multiplexer,* strictly speaking, is a data selector and/or a distribution (see Chapter 10). In general, they all fall under the heading of conversion and switching devices used in digital systems. For example, two relatively inexpensive encoding

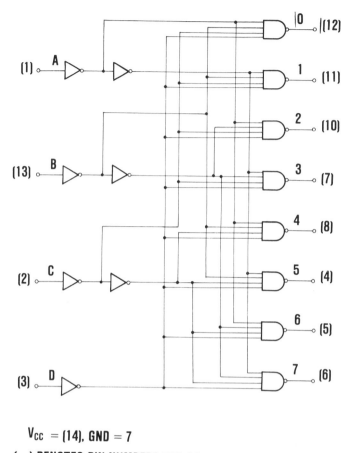

$V_{CC} = (14)$, GND = 7

[ ] DENOTES PIN NUMBERS FOR 14 PIN DUAL-IN-LINE PACKAGE

Figure 9-5: Logic diagram for the 8250 binary-to-octal decoder

systems that come in **IC** form are the 74147 and 74149 10-line to 4-line and 8-line to 3-line priority encoders.

These **TTL** encoders feature *priority decoding* of the inputs so that only the highest order data line is *encoded*. Notice, we have both decoding and encoding taking place in the same **IC**, in this case. But the **IC**'s are definitely encoders. The 74147 encodes nine data lines to 4-line (8-4-2-1) BCD. The 74148 encodes eight data lines to 3-line (4-2-1) binary (octal). As you can see, this fits our definition of an encoder: An uncoded value is converted to a coded form. A functional block diagram of the 74147 is shown in Figure 9-6. The numbers in

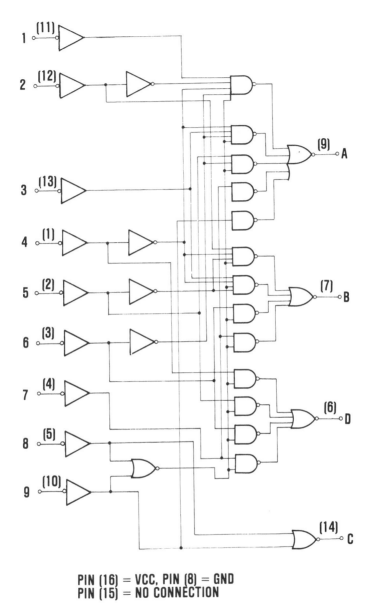

PIN (16) = VCC, PIN (8) = GND
PIN (15) = NO CONNECTION

Figure 9-6: Functional block diagram of a 74147. This IC encodes nine data lines to 4-line (8-4-2-1) BCD. See Figure 9-8 for its truth table

parenthesis are pin numbers of the **IC** (see Figure 9-7). The numbers outside parenthesis and lettered terminals are referring to the function table inputs and outputs (see Table 9-4).

As an example, let's assume that you have a 74147 wired up on a breadboard, with switches on the inputs 1 through 9, and LED's on outputs A through C. Of course V_{cc} (5V) must be applied to pin 16 and ground to pin 8. The actual pin configuration for this **IC** is shown in Figure 9-7.

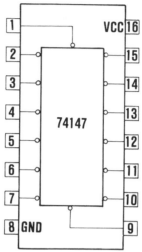

Figure 9-7: Pin configuration for a 74147 that can be used as a guide during breadboard mounting of the IC

After the **IC** is properly mounted and power is applied, your next step is to refer to the function table shown in Table 9-4 to see

INPUTS									OUTPUTS			
1	2	3	4	5	6	7	8	9	D	C	B	A
H	H	H	H	H	H	H	H	H	H	H	H	H
X	X	X	X	X	X	X	X	L	L	H	H	L
X	X	X	X	X	X	X	L	H	L	H	H	H
X	X	X	X	X	X	L	H	H	H	L	L	L
X	X	X	X	X	L	H	H	H	H	L	L	H
X	X	X	X	L	H	H	H	H	H	L	H	L
X	X	X	L	H	H	H	H	H	H	L	H	H
X	X	L	H	H	H	H	H	H	H	H	L	L
X	L	H	H	H	H	H	H	H	H	H	L	H
L	H	H	H	H	H	H	H	H	H	H	H	L

H = HIGH LOGIC LEVEL, L = LOW LOGIC LEVEL
X = IRRELEVANT

Table 9-4: Function table for a 74147

which LED should light when you open or close selected switches on the inputs. First, the implied decimal zero condition requires no input condition, as zero is encoded when all nine data switches are at a high logic level (see the top line - all H's - of the function table). All data inputs and outputs are active at the low logic level.

Next, notice that it is irrevelant what input data is entered on inputs 1 through 8 if 9 is set to a logic low level. The output will be 1001 (LHHL) BCD. In our example, two LED's will be on and two off, i.e., decimal number 9 being 1001. Checking the other lines in the function table, you'll see that the encoder will perform priority decoding of the inputs to ensure that only the highest order data line is encoded. The next line, output LHHH, translates to 1000 binary, or 8 decimal, when input 8 is low and input 9 is high, and each line following can be read in the same manner.

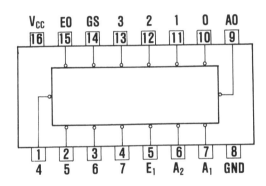

INPUTS									OUTPUTS				
E_1	0	1	2	3	4	5	6	7	A_2	A_1	A_0	GS	EO
H	X	X	X	X	X	X	X	X	H	H	H	H	H
L	H	H	H	H	H	H	H	H	H	H	H	H	L
L	X	X	X	X	X	X	X	L	L	L	L	L	H
L	X	X	X	X	X	X	L	H	L	L	H	L	H
L	X	X	X	X	X	L	H	H	L	H	L	L	H
L	X	X	X	X	L	H	H	H	L	H	H	L	H
L	X	X	X	L	H	H	H	H	H	L	L	L	H
L	X	X	L	H	H	H	H	H	H	L	H	L	H
L	X	L	H	H	H	H	H	H	H	H	L	L	H
L	L	H	H	H	H	H	H	H	H	H	H	L	H

Figure 9-8:   Pin configuration and function table for a 74148 8-line-to-3-line priority encoder

An 8-line-to-3-line-decoder, the 74148, pin configuration and function table are shown in Figure 9-8. There are two pins (labeled E1 and E0) on this **IC** that are not utilized in the same manner as with the 74147. Cascading circuitry (enable input E1 and enable output E0) has been provided to allow octal expansion without the need for external circuitry. As with the 74147, data inputs and outputs are active at the low logic level.

Another priority circuit you may want to experiment with uses an MC14530 *majority logic gate.* Before we look at this **IC**, perhaps we should first explain what a majority gate is. Simply, this gate is a logic element that will produce a logic high on its output if more than half its inputs are at a logic high. On the other hand, the output will be a logic low for other input conditions. *Note:* Actually, the output will follow any given input logic level, high or low. Figure 9-9 shows an example of what we mean. This gate has five inputs and the output is a logic high whenever any three or more of its inputs are at a logic high. It doesn't make any difference which inputs you make high . . . any three you choose will cause the output to go high.

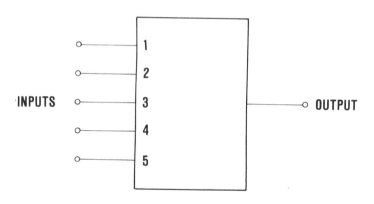

Figure 9-9: Majority logic gate. When the majority of the inputs (1 through 5) assume a given logic level, so does the output

Now, back to the MC14530. This **IC** has two majority logic gates (it is a dual 5-input majority logic gate). Figure 9-10 shows a block diagram of one of the gates and its pin connections. The other half is exactly the same except the input pins are 9, 10, 11, 12, and 13. The input pin to its exclusive **NOR** gate is 14. The output is pin 15.

To produce a logic high on both outputs when a majority of inputs are at a logic high, both pins 6 and 14 (the inputs to both ex-

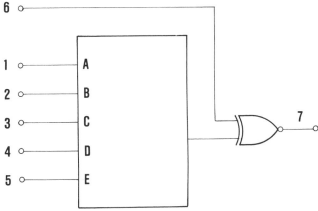

Figure 9-10: One of two 5-input majority logic gates and its NOR gate. The output (pin 7) will be a logic 1 if any three or more inputs are logic 1, and pin 6 is tied to a logic high

clusive **OR** gates) should be tied to a high (V_{cc}). Pin 16 is shown in Figure 9-11. But, if you connect pins 6 and 14 to a logic low (ground),

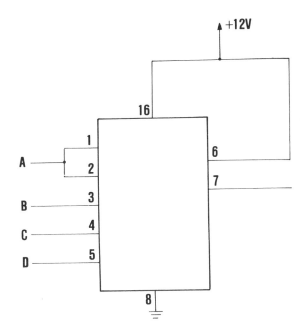

Figure 9-11: Wiring diagram for an MC14530 dual 5-input majority logic gate that will give priority to a selected input

the output will be *low* whenever three or more inputs are high. The circuit shown in Figure 9-11 shows how you can use the **IC** to select a certain input over all the rest. To say it another way, give priority to a certain input.

By connecting any two inputs together, you can make the resulting input switch as important as the other inputs (see Figure 9-11). In our example, if A or any one of the other inputs goes high, it will sound an alarm, light an LED, or whatever you desire. Some other experiments you might like to try with this **IC** are:

1. Place a high on pins 1, 2, and 6 (see Figure 9-10). Now, if you place a high on any other pin, you have a 3-input **OR** gate.

2. Place pin 6 to ground (a logic low). Place pin 1 and 2 at logic high. This will invert the output, as you'll remember. Therefore, you now have a 3-input **NOR** gate.

3. Place pins 1 and 2 at ground. Place pin 6 at a high. You now have a 3-input **AND** gate.

4. Again place pins 1 and 2 at ground, Also pin 6 to ground. You have a 3-input **NAND** gate, in this case.

You can use this same idea with othe **IC**'s, however, *do not short more than one output of any IC during any experiment.* As a general rule, you can short several *input* pins together or to ground, *but not output pins!*

Decoder/Display Drivers

In Chapter 1, we discussed the fact that all 7-segment displays need a decoder in order to operate properly. Typically, you will find these units to be BCD-to-7-segment **IC**'s. There are three standard **TTL** family **IC**'s in use and these are the 7446, 7447, and 7448. The interface **IC**'s, Signetics 8T04/5/6 that we examined in Chapter 1 are a bit different from these (7446/7/8), as we shall see. All three of the 74's have the same pin configuration. See Figure 9-12.

The 1-2-4-8 BCD input is applied to A (pin 7), B (pin 1), C (pin 2), and D (pin 6), respectively. The seven outputs are as shown in Figure 9-12 (pin 9 through 15 are labeled e, d, c, b, a, and g, respec-

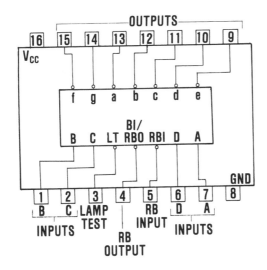

Figure 9-12: Pin configuration for the 7446, 7447, and 7448 BCD-to-7-segment decoder/driver (Courtesy Signetics)

tively). Pin 3, labeled *lamp test,* is a lamp test input that will turn on all seven outputs simultaneously if a logic low is applied to the pin. Also incorporated in these devices is a blanking circuit, allowing leading and trailing zero suppression. This blanking input (RB input, output pins are 4 and 5) will provide the function described by the truth table and notes shown in Table 9-5 on page 240.

The 7448 has resistor pull-up on the outputs, to provide source current to drive interface elements. The source current is about 2 mA and the output sink current is 6.4 mA (max.). This **IC** has active-high outputs and it may require pull-up resistors of about 1000 ohms for proper operation. Another difference between the 7446/7 and the 7448 is found in Note 4, in Table 9-5. *Notice,* Note 4 states: When blanking input/ripple-blocking output is open or held at a logic 1, and a logic 0 is applied to lamp-test input, all segment outputs go to a logic 0. When working with a 7448, and you do exactly as stated, all segment outputs go to a logic 1. Figure 9-13 can be used for segment identification when working with all of these **IC**'s.

Seven-Segment LED Read-Outs

Figure 3-20, in Chapter 3, shows several discrete LED indicator circuits that you can use. However, there are other LED ar-

DECIMAL OR FRACTION	LT	RBI	D	C	B	A	BI/RBO	a	b	c	d	e	f	g	NOTE
	⟵		INPUTS			⟶		⟵			OUTPUTS			⟶	
0	1	1	0	0	0	0	1	0	0	0	0	0	0	1	1
1	1	X	0	0	0	1	1	1	0	0	1	1	1	1	1
2	1	X	0	0	1	0	1	0	0	1	0	0	1	0	
3	1	X	0	0	1	1	1	0	0	0	0	1	1	0	
4	1	X	0	1	0	0	1	1	0	0	1	1	0	0	
5	1	X	0	1	0	1	1	0	1	0	0	1	0	0	
6	1	X	0	1	1	0	1	1	1	0	0	0	0	0	
7	1	X	0	1	1	1	1	0	0	0	1	1	1	1	
8	1	X	1	0	0	0	1	0	0	0	0	0	0	0	
9	1	X	1	0	0	1	1	0	0	0	1	1	0	0	
10	1	X	1	0	1	0	1	1	1	1	0	0	1	0	
11	1	X	1	0	1	1	1	1	1	0	0	1	1	0	
12	1	X	1	1	0	0	1	1	0	1	1	1	0	0	
13	1	X	1	1	0	1	1	0	1	1	0	1	0	0	
14	1	X	1	1	1	0	1	1	1	1	0	0	0	0	
15	1	X	1	1	1	1	1	1	1	1	1	1	1	1	
BI	X	X	X	X	X	X	0	1	1	1	1	1	1	1	2
RBI	1	0	0	0	0	0	0	1	1	1	1	1	1	1	3
LT	0	X	X	X	X	X	1	0	0	0	0	0	0	0	4

NOTES:

1. BI/BRO is wire-OR logic serving as blanking input (BI) and/or ripple-blanking output (RBO). The blanking input must be open or held at a logical 1 when output functions 0 through 15 are desired and ripple-blanking input (RBI) must be open or at a logical 1 during the decimal 0 input. X = input may be high or low.

2. When a logical 0 is applied to the blanking input (forced condition) all segment outputs go to a logical 1 regardless of the state of any other input condition.

3. When ripple-blanking input (RBI) is at a logical 0 and A = B = C = D = logical 1, all segment outputs go to a logical 1 and the ripple-blanking output goes to a logical 0 (response condition).

4. When blanking input/ripple-blanking output is open or held at a logical 1, and a logical 0 is applied to lamp-test input, all segment outputs go to a logical 0.

Table 9-5: Truth table for the 7446 and 7447 BCD-to-7-segment decoder/driver. The truth table for the 7448 is exactly the same, except it is the complement of this table

SEGMENT IDENTIFICATION

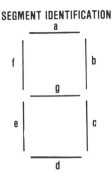

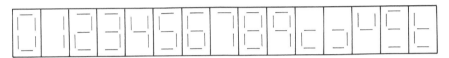

NUMERICAL DESIGNATION-RESULTANT DISPLAYS

Figure 9-13: Segment identification. See Table 9-5 and notes. The various segments needed to produce the ten decimal digits are shown in Figure 1-9

rangements that you will need to know about in your work. These are the *common anode* and *common cathode* LED 7-segment displays that are shown in Figure 9-14.

As with many other electronic components, observation of polarity is important when using LED 7-segment displays. Notice, V_{cc} is connected to the anode of any LED symbol shown in Figure 9-14 (A) (this configuration is called *common anode*). Therefore, this type requires lows (ground) on the inputs, to turn on the segments. The other type — common cathode — requires highs (usually equal to V_{cc}) to turn on its segments. See Figure 9-14 (B). Table 9-6 contains a list of various LED displays with descriptions and pin functions.

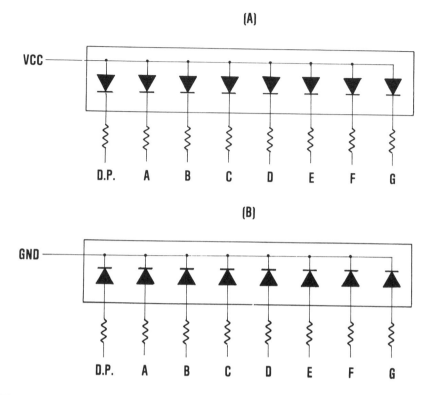

Figure 9-14: LED 7-segment display. (A) common anode, (B) common cathode

Table 9-6

Description and Pin Connections For Various LED Displays

TIL 312 (Texas Inst.) and 7450014 (Dialight). Common anode, color red, 0.3 inch

Pin Function	
No. 1. Cathode A	No. 8. Cathode D
No. 2. Cathode F	No. 9. Cathode RDP
No. 3. Anode digit & D.P.	No. 10. Cathode C
No. 4. Omitted	No. 11. Cathode G
No. 5. Omitted	No. 12. Omitted
No. 6. Cathode L.D.P.	No. 13. Cathode B
No. 7. Cathode E	No. 14. Cathode Digit & D.P.

TIL 313 (Texas Inst.) and 7450016 (Dialight). Common cathode, color red, 0.3

Pin Function

No. 1. Omitted	No. 8. Omitted
No. 2. Cathode*	No. 9. Cathode*
No. 3. Anode F	No. 10. Anode RHDP
No. 4. Anode G	No. 11. Anode C
No. 5. Anode	No. 12. Anode B
No. 6. Andoe D	No. 13. Anode A
No. 7. Omitted	No. 14. Omitted

*Pins 2 and 9 are internally connected

MAN52A (Monsanto), color green, 0.3 inch, MAN3620A (Monsanto), color orange, 0.3 inch, MAN72A (Monsanto), color red, 0.3 inch, MAN82A (Monsanto), color yellow, 0.3 inch

Pin connections

No. 1. Cathode A	No. 8. Cathode D
No. 2. Cathode F	No. 9. _____
No. 3. Anode	No. 10. Cathode C
No. 4. _____	No. 11. Cathode G
No. 5. _____	No. 12. _____
No. 6. Cathode D.P.	No. 13. Cathode B
No. 7. Cathode E	No. 14. Anode

MAN54 (Monsanto), color green, 0.3 inch, MAN3640A (Monsanto), color orange, 0.3 inch, AN74A, (Monsanto). color red, 0.3 inch

All common cathode, common anode

Pin Connections

No. 1. Anode F	No. 8. Anode C
No. 2. Anode G	No. 9. Anode D.P.
No. 3. _____	No. 10. _____
No. 4. Cathode	No. 11. _____
No. 5. _____	No. 12. Cathode
No. 6. Cathode D.P.	No. 13. _____
No. 7. _____	No. 14. Anode

DL747 (Litronix), common anode, color red, 0.63 inch

Pin Connections

No. 1. Cathode A	No. 8. Cathode D
No. 2. Cathode F	No. 9. Anode

No. 3. Anode
No. 4. Cathode E
No. 5. Anode
No. 6. Cathode D.P.
No. 7. _____

No. 10. Cathode C
No. 11. Cathode G
No. 12. Cathode B
No. 13. _____
No. 14. Anode

DL750 (Litronix), common cathode, color red, 0.63 inch

Pin Connections
No. 1. Anode A
No. 2. Anode F
No. 3. Cathode
No. 4. Anode E.
No. 5. Cathode
No. 6. Anode D.P.
No. 7. _____

No. 8. Anode D
No. 9. Cathode
No. 10. Anode C
No. 11. Anode G
No. 12. Anode B
No. 13. _____
No. 14. Cathode

Alphanumeric Display

Two 5 x 7 LED array alphanumeric displays capable of producing numbers, letters and special characters are the TIL305 (Texas Inst.) and 745-0005 (Dialight). A 14-pin DIP exact replacement for either of these 5 x 7 matrix displays (or others of the same type) is shown in Figure 9-15.

Figure 9-15: 5 x 7 array alphanumeric display (0.3 inch, color red)

The LED's in these display DIP's are arranged in a 5 x 7 matrix, where the *rows* are the cathodes and the *columns* are the anodes (see Figure 9-16). Referring to Figure 9-16, if you wish to light the upper left-hand LED in either the TIL305, 745-0005, or similar

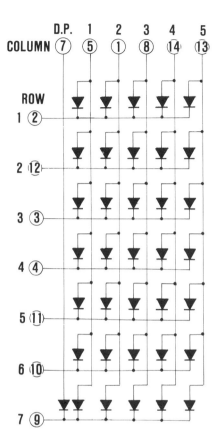

Figure 9-16: 5 x 7 LED array schematic diagram for a TIL305 or similar display

device, row 1 (pin 2) would have to be brought to a low and column 1 (pin 5) would have be brought to a high.

Next, let's assume that you wish to produce a certain letter on one of these displays. In actual practice (such as in a computer), a **RAM** or **ROM** is usually connected to the 7-row inputs to produce the selected character (i.e., letter, etc.) as the columns are scanned. The various rows are generally brought low, while each column is brought high, by the scanning device. But you can produce one letter at a time by using current-limiting resistors in the anode lines.

Place a 330 ohm resistor on pins 7, 5, 1, 8, and 14. Use + 5V as V_{cc}. *Note:* If you use a **TTL IC** to drive any of these displays, it may be necessary to use a smaller value of resistance in order to achieve the brightness you desire. Watch it! Most of these type displays are rated at *20 mA maximum current for each LED* (0.5 mA per segment).

As an example of placing a certain character on the display: Place column 1 (pin 5) at a high, then all rows at a low. This will cause all LED's on the entire left-hand side of the LED display to light, producing one side of the letter H (we will use H for our example). Next, the cross bar of the letter H. When column 2 is high, row 4 is low, column 3 is high, row 4 low, column 4 high, row 4 low. Finally, the right-side of the letter H. Column 5 high, all rows low. The result is shown in Figure 9-17.

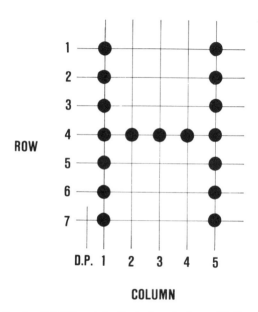

Figure 9-17:   5 x 7 LED matrix displaying the letter H. See Figure 9-16 for a schematic diagram of a 5 x 7 LED matrix

EXPERIMENT 9-1

Test LED Displays

When testing a 7-segment LED display, you should have the manufacturer's data sheets or use the information given in this chapter (if the display is known to be similar or an exact replacement). See

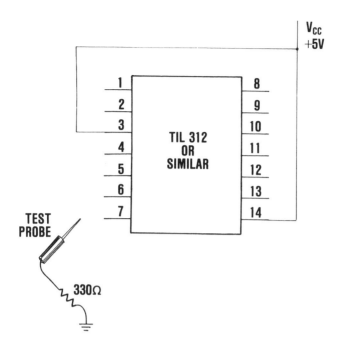

Figure 9-18: Testing a common anode LED display. Connect all anode pins to a recommended V_{CC}. Touch all cathode segment pins with a grounded test probe, as shown

Table 9-6 and/or Figure 9-16. The basic testing of a TIL312 common anode, 7-segment display is shown in Figure 9-18.

By referring to Table 9-6, you will see that the LED display (TIL312) we are using in our example, has seven segment pins, cathode A, B, C, D, E, and F. Touch each of these pins with the test probe and you should see each segment light up. With a TIL313, a common cathode, connect all cathode pins to ground (pins 2 and 9, in this case) and then use the 330 resistor and test probe (after you have connected the open end of the resistor to +5V) to touch all anode segment pins (anodes A through F).

To test the 5 x 7 LED array shown in Figure 9-16 (TIL305 or similar), connect all cathode pins to ground and then touch the test probe and resistor combination (connected to +5V) to each anode pin. When you do this, you should see the LED's in each column shown in Figure 9-16, light up.

EXPERIMENT 9-2

Testing 7-Segment Decoder/Driver IC's

A BCD 7-segment decoder/driver **IC** can be connected to drive the segments of an LED readout, as shown in Figure 9-19. *Note:* The example shown is a common anode display system.

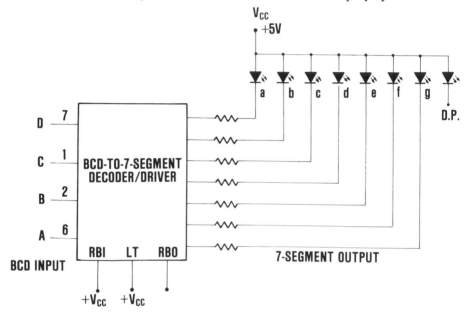

Figure 9-19: Wiring diagram for testing a 7-segment decoder/driver IC using LED readout

Connect the ripple blanking input (used for leading and trailing zero suppression) and lamp test pin to V_{cc}. This places these two inputs at a logic high, required to display numbers 0 through 9, as you change the inputs. For example, Figure 9-12 shows the pin configuration for a 7447 BCD-to-7-segment decoder/driver that can be used for the decoder/driver shown in Figure 9-19. The ripple blanking input pin is 5, and the lamp test pin is 3. This **IC** is called a *decoder/driver* because its outputs are buffered to a voltage rating of 15V and a current of 20 mA. Also, you will find a truth table in Table 9-5 that you can use to check out the 7447. Seven-segment decoders are available with different output voltage and current ratings. Moreover, devices with complemented outputs are available to drive common cathode

LED readouts. However, when working with an **IC** such as a 7447, your BCD input pins are A, B, C, and D and, of course, you must connect pin 16 (V_{cc}) to 5 V and pin 8 to ground before testing the device.

Now, because this is a common anode display, you know that a certain output of the decoder/driver must be low in order to turn on a selected LED. As an example, if you place the input to the decoder/driver **IC** at 1001 (high, low, low, high), what should you see on the LED readout? First, when you choose 1001 BCD, you select the decimal number 9. When your input is 9, the truth table shown in Table 9-5 shows that the output should be:

$$abcdefg = 0001100$$

The LED display reads number 9. To prove this to yourself, remember that each *low* turns on a segment in the display. Next, referring to the segment identification shown in Figure 9-20, you will see each segment that is turned on is drawn with a darker line. The on segments are g, f, c, b, and a. The off are d and e (see truth Table 9-5).

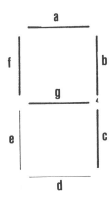

Figure 9-20: Segment identification showing that with a common anode LED display, the input 0001100 to the LED display will produce a decimal number 9 on that display

Figure 9-21 shows a BCD input, 7-segment letters and decimal number display. *Note:* With multi-digit displays, you will need a multiplex system with clock and sequencer.

Referring to Figure 6-1 (Chapter 6), you will find it to be the pin configuration for a 7490 decode counter (the truth table is shown

BCD				7-SEGMENT							DISPLAY
D	C	B	A	a	b	c	d	e	f	g	
0	0	0	0	0	0	0	0	0	0	1	
0	0	0	1	1	0	0	1	1	1	1	
0	0	1	0	0	0	1	0	0	1	0	
0	0	1	1	0	0	0	0	1	1	0	
0	1	0	0	1	0	0	1	1	0	0	
0	1	0	1	0	1	0	0	1	0	0	
0	1	1	0	1	1	0	0	0	0	0	
0	1	1	1	0	0	0	1	1	1	1	
0	0	0	0	0	0	0	0	0	0	0	
1	0	0	1	0	0	0	1	1	0	0	

Figure 9-21: Seven-segment code for experiment 9-2

in Table 6-1). Connect this **IC**'s output pins to the input pins of the 7447 shown in Figure 9-12 and then wire up the LED display shown in Figure 9-19. This setup should produce the entire count on your LED readout. Figure 9-22 shows the basic wiring diagram for this test.

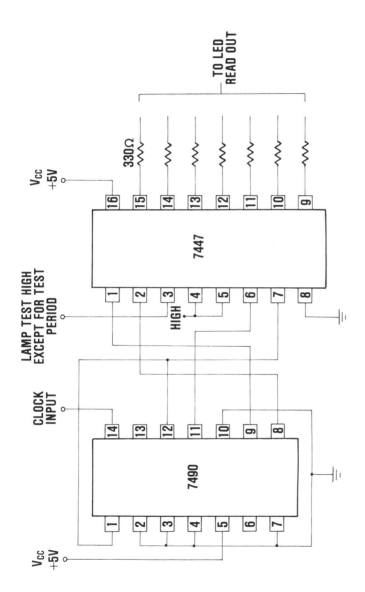

Figure 9-22: Basic wiring diagram for connecting a decoder counter IC, 7-segment decoder/driver IC and LED readout, to display decimal numbers

CHAPTER 10

Practical Applications Using
Data Selector Multiplexers
And Demultiplexers

This chapter provides an easy-to-understand approach to **IC**'s that you can use in digital systems for coordinating the operations of two or more subsystems mounted on your breadboard, in a computer, or similar application. As the name implies, a *data selector/multiplexer* is a circuit that is capable of looking at two or more digital inputs and selecting one of the inputs, and passing that input on to the output. The following pages provide both technicians and experimenters with step-by-step instructions on how to use these versatile devices.

Understanding Digital Multiplexers

An obvious use of a multiplexer is that of moving data from one group of digital circuits (such as registers) to a common ouput bus line. It is important that you realize that the whole point of using a multiplexer in this manner is to make data transmission more efficient by replacing a large number of data channels on your breadboard, etc., with a single *time-multiplexed channel.* At this point, you may be wondering how you will ever recover the data once it is time-multiplexed. Later in this chapter, you will find a section on demultiplexer circuits, that will answer your questions on this subject.

Perhaps the best way to understand a multiplexer is to examine an **IC** such as the **TTL 74157** (4019 **CMOS**) quadruple 2-input data selector/multiplexer. Incidentally, the 74158 is identical to the 74157 and 4019 except its outputs are inverted. One way to look at these devices is as logical implementations of a 4-pole 2-position switch, with the position of the switch being set by logic levels supplied to a select input. Figure 10-1 shows an example of a mechanical switch of this type.

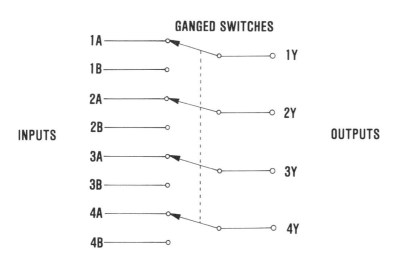

GANGED SWITCHES

INPUTS

OUTPUTS

Figure 10-1: Hand-operated switch that is analogous to a 74157 2-input data selector/multiplexer

The mechanical switch shown in Figure 10-1 is, of course, operated by hand to connect selected input lines to the output lines. The 74157 digital multiplexer will perform the same operation when the appropriate logic level is applied to the select and strobe inputs. See Figure 10-2 for a logic diagram of an actual multiplexer, the 74157 or 74158.

Now, by referring to Figures 10-1 and 10-2, we can see how this device will perform the function of a 2-input data selector. Notice, the top circuit can select data from two sources, 1A, and 1B,

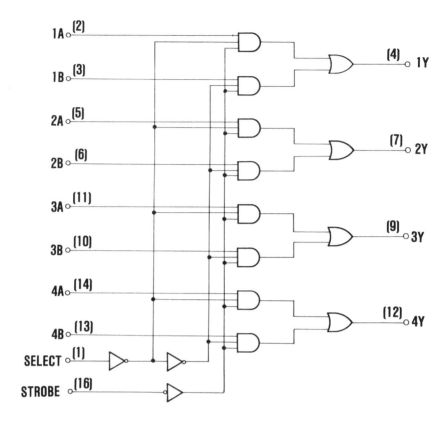

Figure 10-2: Logic diagram for a 74157 quadruple 2-input data selector/multiplexer

and this data from either source is fed out to one output, 1Y. Therefore, we can call this a *2-to-1 multiplexer*. Or, to put it another way, this section reduces two lines of incoming data to a single output line. Switch the switch S1 back and forth in a time sequence and we have a time-multiplexed output.

By referring to the logic diagram shown in Figure 10-2, you will see that there are actually four of these 2-to-1 multiplexers in the 74157 **IC**, therefore, this **IC** is called a *quadruple 2-to-1 multiplexer*. Table 10-1 shows a truth table for the 74154. Notice, setting the select input to a logic low makes the A inputs appear at the Y outputs. Setting the select input to a logic high, directs the B inputs to the Y outputs. The **IC** is, thus, capable of directing one of two input 4-bit binary words to the Y outputs, depending on which logic level you choose to place on the select input. *Note:* The *strobe input* is shown in Table 10-1 and Figure 10-2. According to the truth table, the strobe input is capable of turning off the multiplexer action altogether. In fact, place a logic high on this input and you should find a logic low on the Y outputs, regardless of what you place on the A, B, or select inputs. This strobe feature is especially useful for expanding multiplexer functions; that is, building up to an 8-to-1, 16-to-1, or anything similar, multiplexer.

INPUTS			OUTPUT
STROBE	SELECT	A B	Y
H	X	X X	L
L	L	L X	L
L	L	H X	H
L	H	X L	L
L	H	X H	H

Table 10-1:   Truth table for a 74157 quad 2-input-to-1-output data selector/multiplexer. Table is for one of the four sections

A Guide to Using Demulitplexers

As the name suggests, a demultiplexer is the opposite of a multiplexer. That is, it is a switching device that can select one input and connect it to one output line of two or more outputs. Figure 10-3 shows a comparison of switch-equivalents for multiplexers and demultiplexers.

Figure 10-3:   A comparison of switch-equivalents for multiplexers and demultiplexers. (A) is a 2-to-1 multiplexer equivalent and (B) is a 1-to-2 demultiplexer equivalent

When you are using a multiplex circuit such as the switch-equivalent shown in Figure 10-3 (A), it is interesting to note that the device is converting parallel data into one line of serial or time-multiplexed data, as we have said. But, when the switching circuit is used as a data demultiplexer, it is converting one line of serial or time-multiplexed data back into parallel data.

If you look up the specs for a 74155 or 74156, you'll find they are usually referred to as *dual 2-line-to-4-line decoder/1-line-to-4-line demultiplexers*. This would seem to indicate that a demultiplexer can be used as a decoder, and this is true. However, it is the job you assign to the **IC** that makes the difference, rather than the **IC** you choose to use. In other words, you usually don't wire the circuit one way for demultiplexing and another way for decoding. Figure 10-4 shows the pin configuration for a 74155. Table 10-2 is a truth table for this **IC**.

When the 74155 is used as a data demultiplexer (to convert time-multiplexed data back to parallel data), you are interested in the number of data inputs and data outputs; for instance, a 1-to-4-demultiplexer, etc. But, if you are using the **IC** as a data selector,

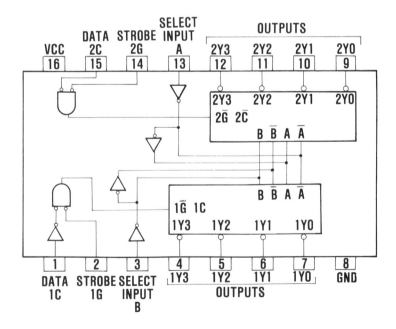

Figure 10-4: Pin configuration for a 74155/74156 dual 2-line-to-4-line decoder/demultiplexer

(A) 2-LINE TO 4-LINE DECODER OR 1-LINE TO 4-LINE DEMULTIPLEXER

SELECT		STROBE	DATA				
INPUTS				OUTPUTS			
B	A	1G	1C	1Y0	1Y1	1Y2	1Y3
X	X	H	X	H	H	H	H
L	L	L	H	L	H	H	H
L	H	L	H	H	L	H	H
H	L	L	H	H	H	L	H
H	H	L	H	H	H	H	L
X	X	X	L	H	H	H	H

SELECT		STROBE	DATA				
INPUTS				OUTPUTS			
B	A	2G	2C	2Y0	2Y1	2Y2	2Y3
X	X	H	X	H	H	H	H
L	L	L	L	L	H	H	H
L	H	L	L	H	L	H	H
H	L	L	L	H	H	L	H
H	H	L	L	H	H	H	L
X	X	X	H	H	H	H	H

(B) 3-LINE TO 8-LINE DECODER TO 8-LINE TO 1-LINE DEMULTIPLEXER

SELECT			STROBE OR DATA	OUTPUTS							
				(0)	(1)	(2)	(3)	(4)	(5)	(6)	(7)
C	B	A	G	2Y0	2Y1	2Y2	2Y3	1Y0	1Y1	1Y2	1Y3
X	X	X	H	H	H	H	H	H	H	H	H
L	L	L	L	L	H	H	H	H	H	H	H
L	L	H	L	H	L	H	H	H	H	H	H
L	H	L	L	H	H	L	H	H	H	H	H
L	H	H	L	H	H	H	L	H	H	H	H
H	L	L	L	H	H	H	H	L	H	H	H
H	L	H	L	H	H	H	H	H	L	H	H
H	H	L	L	H	H	H	H	H	H	L	H
H	H	H	L	H	H	H	H	H	H	H	L

Table 10-2:   Truth table for a 74155/74156

your interest is in the way the outputs will drop from a logic high to a logic low in sequence, one at a time. This is shown in Table 10-2 as a diagonal line of logic lows from top left to bottom right of the truth table. Again, refer to Table 10-2 and you will see that one of the tables has a heading of 3-line-to-8-line decoder and a 1-to-8-demultiplexer. Which of these tables you use will depend on whether you are using the **IC** as a decoder or demultiplexer.

If you want to use this **IC** as a decoder, it will do a very good job as a binary-to-octal converter. Setting the strobe/data input to a logic high lets the three binary inputs (select C, B, and A) select one of the eight outputs (2Y0, 2Y1, 2Y2, 2Y3, and 1Y0, 1Y1, 1Y2, 1Y3). If select A is considered the least significant input bit, 2Y0 could be the octal 0 output, 2Y1 octal 1, 2Y2 octal 2 output and so on. But, as we have said, you also can use the same **IC** as a 1-to-8 demultiplexer. To do this, you must apply the serial data you want to demultiplex to the strobe/data input and you'll have to connect a modulo-8 counter to select C, B, and A inputs.

The truth table in Table 10-2 (B) shows that when the first bit of serial input arrives at the select/data input, for example, the counter should be in a LLL (000) condition to deliver the first bit to the output 2Y0. The next line down, LLH, shows that the counter must deliver a 001 to produce a logic high on 2Y0 and move the logic low over to 2Y1 (drop down in sequence). As you can see, the counter is effectively scanning the input and transmitting the data on to its proper parallel position at the output.

Like most electronics systems, of course, if you connect a counter and demultiplexer in the same circuit they must be synchronized so that the counter is operating in step with the demultiplexer. If they are not synchronized, the truth table becomes useless. Generally, a clock synchronizes the counter so that the selected input line to the multiplexer matches the selected output line of the demultiplexer. Although the notes on the truth table shown in Table 10-2 indicate inputs 1C, 2C, and 1G, 2G are connected together, this is only necessary if you want to perform every function given in the entire truth table. Of course, the individual strobes permit activating or inhibiting each of the 4-bit sections as desired.

Testing Single Sections of a
Dual Decoder/Demultiplexer IC

When testing a single section of an **IC** such as the 74155 or 74156, the 2-line code is applied to select inputs A and B (see Figure 10-4). The 4-line output section (for example, 1Y0, 1Y1, 1Y2, 1Y3) is enabled when you make strobe 1G low and input 1C high. If you wish to test the other 4-line output section (2Y0, 2Y1, 2Y2, 2Y3), enable it by taking both strobe 2G and input 2C low.

These two **IC**'s differ only in that the 74155 has a totem-pole output and the 74156 has an open-collector output. However, you should note that both these **IC**'s share a common set of select inputs, A and B. What this means is that the two sections are selected in

DUAL 2-LINE TO 4-LINE DECODER/1-TO-4-LINE DEMULTIPLEXER

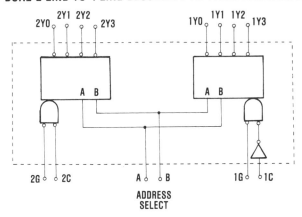

3-LINE TO 8-LINE DECODER/1-TO 8-LINE DEMULTIPLEXER

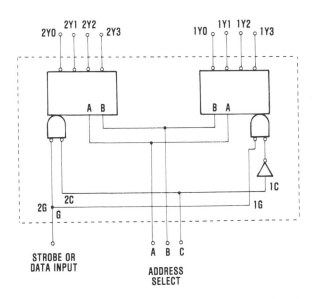

Figure 10-5: Typical applications using 74155 and 74156 decoder/demultiplexers

parallel. But each section does have different data inputs, strobe inputs, and data outputs, as you have seen. Another point worth repeating is that when these **IC**'s are used as a demultiplexer, you use the common strobe line as the data input. Figure 10-5 shows typical applications of these **IC**'s.

A Guide to Multiplexed Digital Displays

The basic multiplexed digital display system shown in Figure 10-6 is an example of how multiplexers and demultiplexers can be used to control data flowing through a digital system. The BCD inputs to the 74153 are to pins, 3, 4, 5, and 6 (to one-half of the multiplexer) and pins 10, 11, 12, and 13 (the other half of the dual 4-to-1 multiplexer). These two multiplexers are controlled (addressed one after the other) by the 7493 **IC**, a 2-bit modulo-4-binary counter.

The select lines address the two multiplexers at the same time. What all this means is that one of the multiplexers is always producing an output of one of the BCD words. For instance, if the counter has a low low (00) output, a BCD digit will appear on the multiplex outputs. Next, when the counter changes its output to low high (01), you should see another input digit appear at the multiplexer outputs.

These examples (and all following BCD inputs) are fed into the 7447's inputs, as shown, and will be processed by the BCD-to-7-segment converter for display on the LED display assemblies, appearing one at a time and in sequence. This may seem to be a strange statement when you consider that all multiplexer select lines are tied together and all LED display assemblies are also connected together. You may wonder how the BCD inputs are made to appear in sequence, under these conditions. To answer this question, take another look at the 7493 counter output lines. Notice, the counter is also addressing a 74156 demultiplexer. If you think about the 74156 as a 2-line-to-4-line decoder, you will quickly realize that the outputs of the **IC** are normally at a logic high, drop to a logic low one at a time and in sequence as the counter output changes the select inputs. Table 10-2 shows the process quite clearly.

Next, these constantly changing select line inputs to the 74156 are then produced on the **IC** outputs and used as inputs to the LED digital assembly driver transistors. These PNP transistors are thus switched on one at a time. As you can see by referring to Figure 10-6,

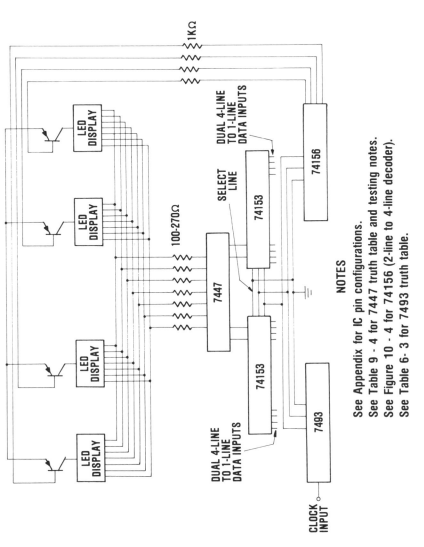

NOTES

See Appendix for IC pin configurations.
See Table 9 - 4 for 7447 truth table and testing notes.
See Figure 10 - 4 for 74156 (2-line to 4-line decoder).
See Table 6- 3 for 7493 truth table.

Figure 10-6: Multiplexing four LED displays

as each transistor is turned on it will cause one of the LED's to show a number, depending on what BCD code number is being fed out of the 7447 **IC**.

If you have trouble seeing how the transistors are turned off and on, remember that a plus voltage on a PNP transistor base will not permit collector current to flow . . . a logic high (in this case) is a plus voltage. At this point, you might think it would be much simpler to use the BCD decoder and display assemblies in the manner explained in Chapter 9. But it is necessary to realize that you must use one BCD-to-7-segment decoder **IC** for each LED readout. With the multiplexed system shown in Figure 10-6, you need only one BCD-to-7-segment decoder **IC** to drive four LED readouts. This becomes particularly important when designing readouts requiring numerous LED digital display assemblies. However, when working with eight digital display assemblies, either method, multiplexed or non-

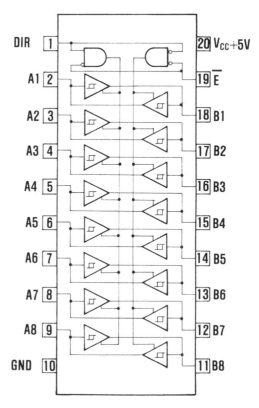

Figure 10-7: Pin configuration and logic symbol for a 74LS645 octal transceiver (Courtesy Monolithic Memories)

multiplexed, comes out about the same so far as the number of **IC**'s required. But it will actually cost you less money to use a non-multiplexed system if you only want to use a 4-digit display. This is because you can get by with only four **IC**'s (four BCD-to-7-segment decoders) and four 7-segment LED displays. With a multiplexed system, you would need at least six **IC**'s (two dual 4-to-1 multiplexers, one BCD-to-7-segment converter, one demultiplexer, one digit driver IC, and a modulo-4 counter) if you are setting up the circuit shown in Figure 10-6.

For the experimenter/designer, there is one very important point if power drain is critical. For example, a typical 1.0-inch-high LED display will draw 20 mA at 3.3 volts. This isn't so bad if you only turn on one digital display at a time, and this is what a multiplexed system does, as we have explained. But, let's say that you are building an 8-digit display and using a non-multiplexed system. The current drain, in this case, is 20 × 8, or 160 mA. Clearly, the mulitplexed system has a great advantage when looking for the lowest possible power drain. When you are building, remember this. A multiplexed display system will not draw more than a single LED display's rated current at any time, due to the fact there is never more than one digital display on at any time.

Using Octal Transceiver IC's for
Two-Way Data Transmission

An interesting octal transceiver that we can use as an example of this type **IC** is the 74LS645. You can use this **IC** in many types of digital systems, for asynchronous two-way communication between data buses. The control is simple and you can switch the direction of all eight data lines, i.e., transmission from A lines to B lines, or from B lines to A lines. The pin configuration and data symbol are shown in Figure 10-7.

The direction of data flow depends on what logic level you place on the direction control (DIR) input (see Table 10-3). Also, by referring to Table 10-3 and Figure 10-7, you will see that the **IC** has an enable input ($\bar{E}$). You can use this input to disable the device so that the buses (all A and B pins) are effectively isolated. A data-to-output delay test setup is shown in Figure 10-8. Typically, you should find that from either A to B, or B to A, is 8 nano seconds.

ENABLE $\overline{E}$	DIRECTION CONTROL DIR	OPERATION
L	L	B DATA TO A BUS
L	H	A DATA TO B BUS
H	X	ISOLATED

Table 10-3: Function table for a 74LS645 octal transceiver. (Courtesy Monolithic Memories)

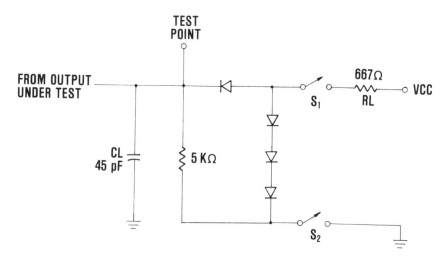

CL INCLUDES PROBE AND
JIG CAPACITORS
ALL DIODES ARE IN916 OR
IN3064

Figure 10-8: Test load for a data to output delay test of a 74LS645 octal transceiver

EXPERIMENT 10-1

Wiring and Test Setup for a Mulitplexer

The 74153 dual 4-line-to-1-line data selector/multiplexer seems to be plentiful and inexpensive, so we will use it for this experiment. Of course, you can also use the same procedure to check other multiplexers such as the 74151 8-line-to-1-line. But you will have to use double the amount of input switches, which is one of the reasons for selecting the 74153 for a first attempt at testing one of these devices. Referring to Figure 10-9, you will see that separate strobe inputs (pins 1 and 15) are provided for each of the two 4-line sections.

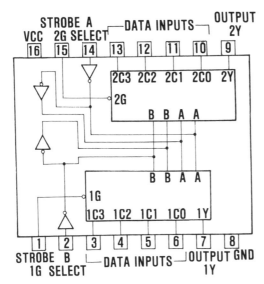

Figure 10-9: Pin configuration for a 74153 dual 4-line-to-1-line data selector/mulitplexer. High level output voltage, typ. = 3.1V, low level, typ. = 0.2V, max. = 0.4V, V_{cc} = 5V

If you place a high-level input on either, or both of the strobe inputs (assuming you have tied these inputs together), you should find a low-level output at pins 7 and 9 (outputs 1Y and 2Y), regardless of what is placed on the address inputs (pins 2 and 14 select) and the data inputs C0 - C3. Therefore, it will require a low-level input on these pins (tie the two pins 1 and 15 together if you want both sections of the IC to be activated at the same time) for this experiment. To select a desired address, connect the inputs A and B to switches and set each switch to a low or high level, using the truth table shown in Table 10-4.

Your next step is to set the data input switches. Set these as desired, using the truth table C0, C1, C2, C3 columns. Then activate the strobe switch and you should see an output on output pin Y. The easiest way to read the output is to connect two LED's to output pins 7 and 8 (1Y and 2Y). Now, repeat these settings for each address.

ADDRESS INPUTS		DATA INPUTS				STROBE	OUTPUT
B	A	C0	C1	C2	C3	G	Y
X	X	X	X	X	X	H	L
L	L	L	X	X	X	L	L
L	L	H	X	X	X	L	H
L	H	X	L	X	X	L	L
L	H	X	H	X	X	L	H
H	L	X	X	L	X	L	L
H	L	X	X	H	X	L	H
H	H	X	X	X	L	L	L
H	H	X	X	X	H	L	H

ADDRESS INPUTS ARE COMMON TO BOTH SECTIONS.
H = HIGH LEVEL, L = LOW LEVEL, X = IRRELEVANT

Table 10-4:   Truth table for the 74153 multiplexer used in Experiment 10-1

Referring to Table 10-4, you'll see that there are nine possible address inputs for A and B.

After you have completed one section of the 74153, test the other section by connecting the data input switches to the other data input pins (for example, first do pins 3, 4, 5, and 6 and then pins 10, 11, 12, and 13) and then repeat the preceding steps, i.e., select the desired address, set the data inputs and press the strobe switch. Do this for each address.

EXPERIMENT 10-2

Wiring and Testing a Decoder/Demultiplexer

Referring to Figure 10-4 and Table 10-2, you will find the pin configuration and truth tables for the 74155 and 74156 dual 2-line-to-4-line decoder/demultiplexer. The difference between these two IC's is that the 74155 has totem-pole outputs and the 74156 utilizes open-collector outputs. Supply voltage for both is typically +5V, min. is 4.75V, max. is 5.25V. For direct drive of LED readouts, use the

74155 (see Note 8, Appendix C). The rest of this experiment, using this **IC**, follows:

Step 1. Tie the two strobe input pins 2 and 14 together and use a toggle switch input to place at a low level

Step 2. Connect switches on the address lines, pins 3 and 13, labeled *select*. Select an address (use Table 10-2).

Step 3. Connect switches on the data inputs, pins 1 and 15 (1C and 2C).

Step 4. Connect LED's on output pins 4, 5, 6, and 7 (1Y3, 1Y2, 1Y1, 1Y0).

Step 5. Place switch on pin 1 (data 1C) to a high and the LED connected to the output you have selected should turn off.

Step 6. Repeat all steps for each address.

Step 7. Check the other section. To do this, place the switch on input 2C (pin 15) to a high level. You should see an LED turn on.

Step 8. Repeat all steps and check each address while checking outputs 2Y3, 2Y2, 2Y1, and 2Y0.

APPENDIX A

Pin configurations for some of the most popular TTL family digital **IC's**. All packages are dual in-line unless noted (Molded or Cerdip).

7400
QUAD 2-INPUT POSITIVE NAND GATES

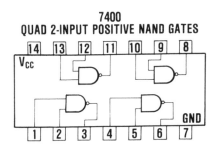

7404
HEX INVERTER

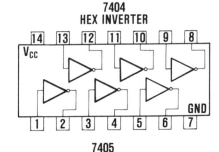

7401
QUAD 2-INPUT POSITIVE NAND GATES
WITH OPEN COLLECTOR OUTPUTS

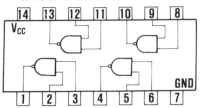

7405
HEX INVERTER WITH OPEN COLLECTOR
OUTPUTS (SEE 7404 FOR PIN CONFIGURATION)

7406
HEX INVERTER BUFFER/DRIVERS WITH
OPEN COLLECTOR OUTPUTS (SEE 7404
FOR PIN CONFIGURATION)

7407
HEX BUFFER/DRIVER WITH OPEN
COLLECTOR HIGH VOLTAGE OUTPUTS

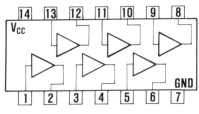

7402
QUAD 2-INPUT POSITIVE NOR GATES

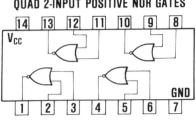

7408
QUADRUPLE 2-INPUT POSITIVE AND GATE

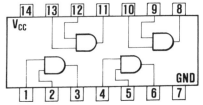

7403
QUAD 2-INPUT POSITIVE NAND GATES
WITH OPEN COLLECTOR OUTPUTS
(SEE 7400 FOR PIN CONFIGURATION)

7409
QUAD 2-INPUT WITH OPEN COLLECTOR
OUTPUTS (SEE 7408 FOR PIN CONFIGURATION)

7410
TRIPLE 3-INPUT POSITIVE NAND GATES

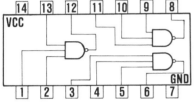

7411
TRIPLE 3-INPUT POSITIVE AND GATES
(SEE 7410 FOR PIN CONFIGURATION)

7413
DUAL 4-INPUT POSITIVE NAND SCHMITT
TRIGGER

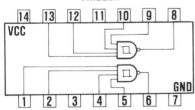

7414
HEX SCHMITT TRIGGER INVERTER

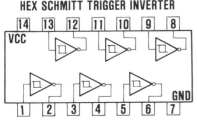

7417
HEX BUFFER/DRIVER WITH OPEN COLLECTOR
HIGH VOLTAGE OUTPUTS (SEE 7414
FOR PIN CONFIGURATION. NOTE: OUTPUTS
ARE NOT INVERTED IN THIS IC)

7420
DUAL 4-INPUT POSITIVE NAND GATES

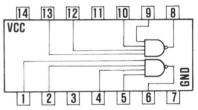

7425
DUAL 4-INPUT POSITIVE NOR GATE WITH
STROBE

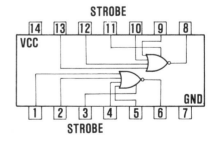

7426
QUAD 2-INPUT HIGH VOLTAGE NAND
GATES (SEE 7400 FOR PIN CONFIGURATION)

7427
TRIPLE 3-INPUT POSITIVE NOR
GATES. (SEE 7410 FOR PIN
CONFIGURATION)

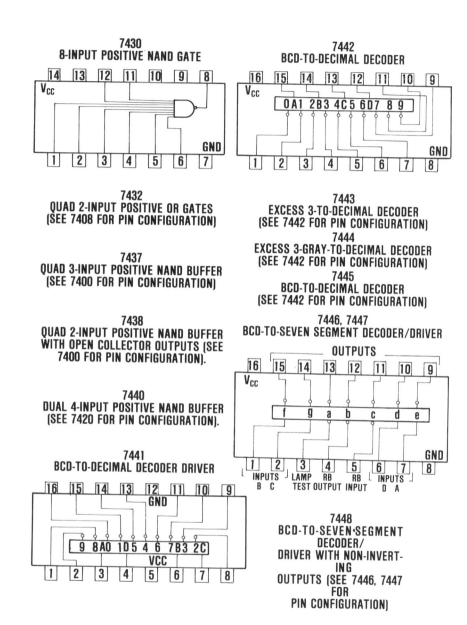

7430
8-INPUT POSITIVE NAND GATE

7432
QUAD 2-INPUT POSITIVE OR GATES
(SEE 7408 FOR PIN CONFIGURATION)

7437
QUAD 3-INPUT POSITIVE NAND BUFFER
(SEE 7400 FOR PIN CONFIGURATION)

7438
QUAD 2-INPUT POSITIVE NAND BUFFER WITH OPEN COLLECTOR OUTPUTS (SEE 7400 FOR PIN CONFIGURATION).

7440
DUAL 4-INPUT POSITIVE NAND BUFFER (SEE 7420 FOR PIN CONFIGURATION).

7441
BCD-TO-DECIMAL DECODER DRIVER

7442
BCD-TO-DECIMAL DECODER

7443
EXCESS 3-TO-DECIMAL DECODER
(SEE 7442 FOR PIN CONFIGURATION)
7444
EXCESS 3-GRAY-TO-DECIMAL DECODER
(SEE 7442 FOR PIN CONFIGURATION)
7445
BCD-TO-DECIMAL DECODER
(SEE 7442 FOR PIN CONFIGURATION)

7446, 7447
BCD-TO-SEVEN SEGMENT DECODER/DRIVER

7448
BCD-TO-SEVEN-SEGMENT DECODER/ DRIVER WITH NON-INVERT-ING OUTPUTS (SEE 7446, 7447 FOR PIN CONFIGURATION)

7450/7451
DUAL 2-SIDE, 2-INPUT AND-OR-INVERT GATES

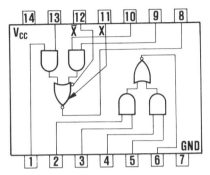

7453/7454
4-WIDE 2-INPUT AND-OR-INVERT GATES

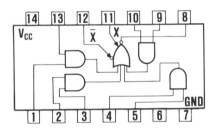

7460
DUAL 4-INPUT EXPANDER

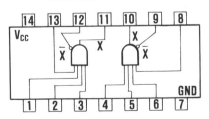

7472
AND GATES J-K FLIP FLOP

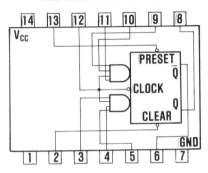

7473
DUAL J-K FLIP-FLOPS

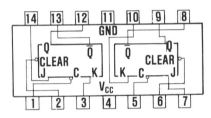

7474
DUAL D-TYPE FLIP-FLOPS

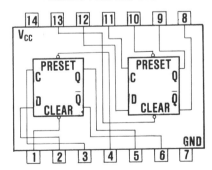

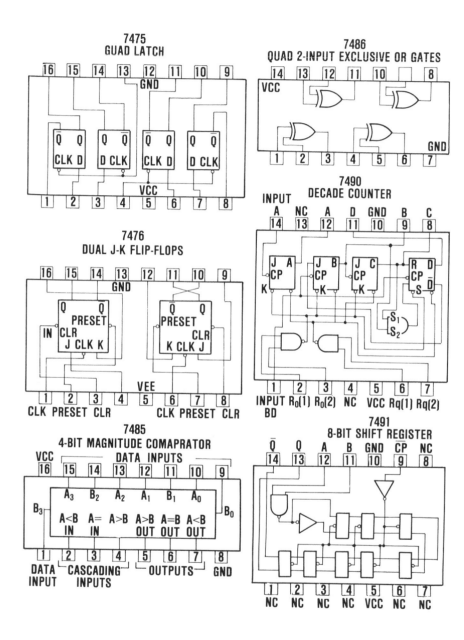

7475
GUAD LATCH

7486
QUAD 2-INPUT EXCLUSIVE OR GATES

7476
DUAL J-K FLIP-FLOPS

7490
DECADE COUNTER

7485
4-BIT MAGNITUDE COMAPRATOR

7491
8-BIT SHIFT REGISTER

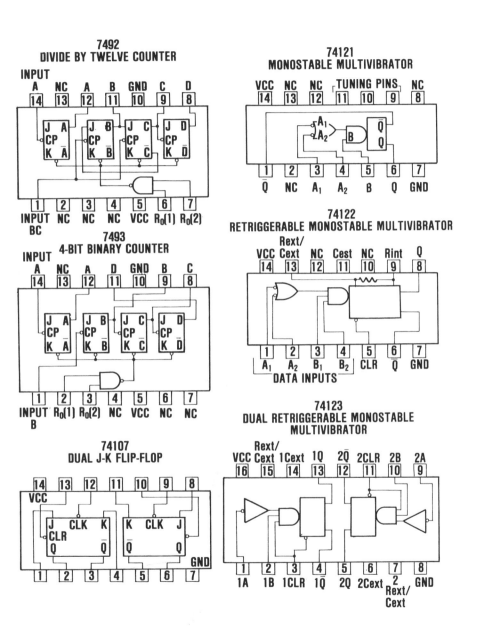

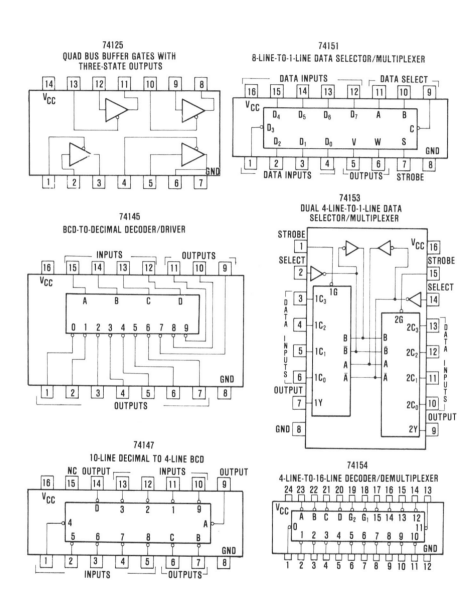

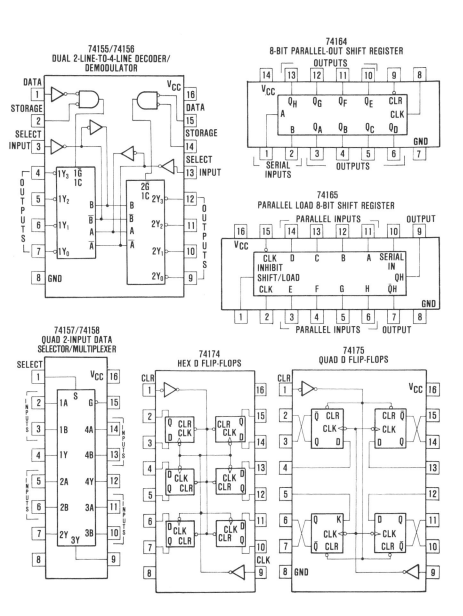

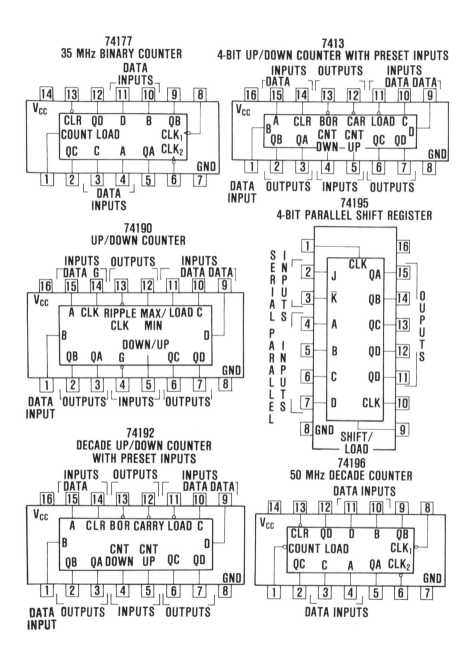

74177
35 MHz BINARY COUNTER

74413
4-BIT UP/DOWN COUNTER WITH PRESET INPUTS

74190
UP/DOWN COUNTER

74195
4-BIT PARALLEL SHIFT REGISTER

74192
DECADE UP/DOWN COUNTER WITH PRESET INPUTS

74196
50 MHz DECADE COUNTER

APPENDIX B

Pin outlines and internal block diagrams of the programmable array logic family, courtesy Monolithic Memories, 1165 E. Arques, Sunnyvale, CA. 94086. This company is the major supplier and developer of these devices, commonly called PAL's®. When reading PAL's identifying code, for example, a PAL14L4, you can use this key:

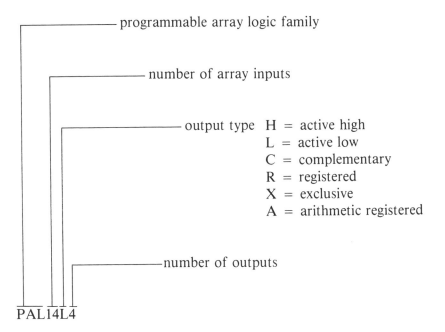

programmable array logic family

number of array inputs

output type   H = active high
              L = active low
              C = complementary
              R = registered
              X = exclusive
              A = arithmetic registered

number of outputs

PAL14L4

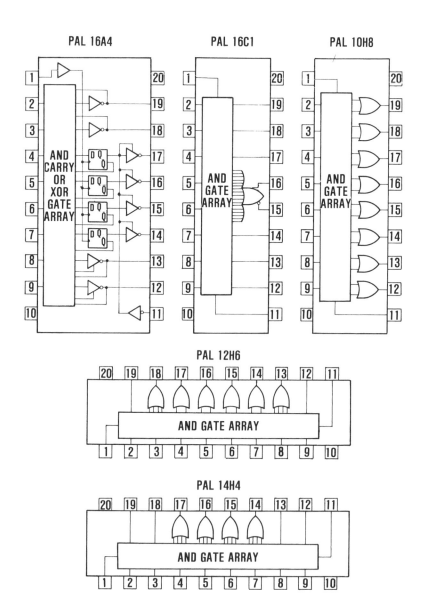

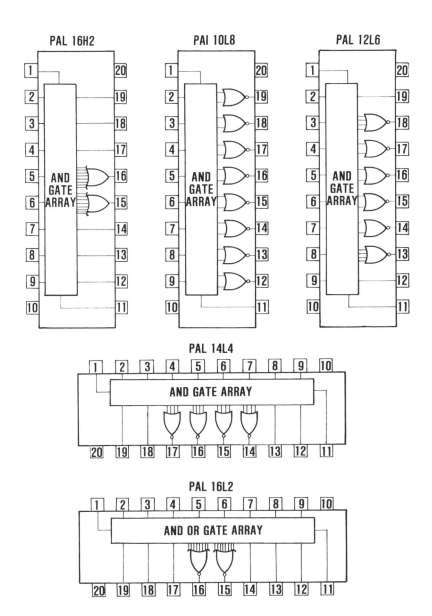

PAL 16L8

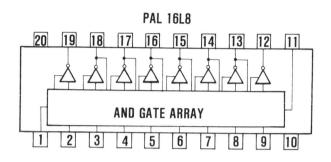

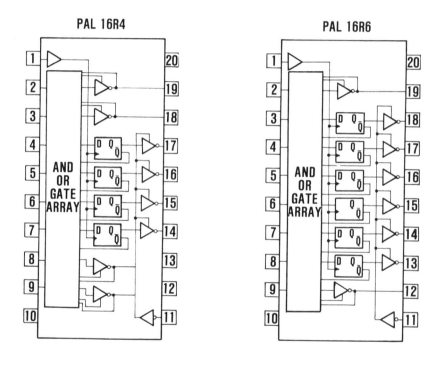

PAL 16R4

PAL 16R6

PAL 16R8

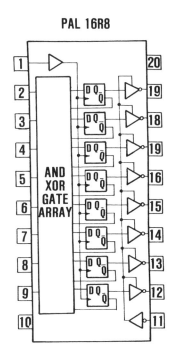

PAL 16X4

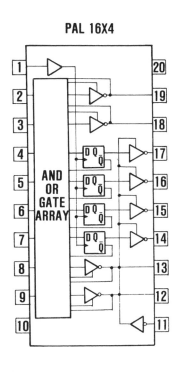

APPENDIX C

List of Critical TTL and
CMOS Use Procedures

Even though **TTL** chips are relatively easy to use and the following list of tricks and procedures have been explained throughout this book, it is extremely important that you review them from time-to-time (particularly if you are new to the field of digital electronics).

1. When working with inputs, do not let the input voltage exceed the supply voltage (generally, +5.0V) and do not permit the input to go below ground potential (less than 0 volts).

2. When selecting a **TTL** circuit power supply, always choose a *regulated supply*. In almost every case, **TTL IC**'s are guaranteed by the manufacturer only if V_{CC} is operated between +4.75 and +5.25 volts; i.e., 5% of a nominal +5V. CMOS voltage range can be from +3 to +15 volts.

3. As a rule of thumb, use number 20 AWG wire for all **TTL** power connections. Also, use a large electrolytic bypass capacitor (4 to 100 μF) between your input power lines and ground.

4. If the **IC** has a totem-pole output circuit (most **TTL IC**'s do), place a 0.01 μF capacitor between the **IC**'s supply input (V_{CC}) and ground. Keep your capacitor leads short and mount it from the **IC**'s V_{CC} pin to ground, if possible. However, if more than one **IC** is connected to a common bus line, a single capacitor can be used. Connect the capacitor between the common +5V line and ground. The purpose of this bypass capacitor is to absorb the large current spikes created by the switching action of the totem-pole output circuit.

5. Keep all wires that carry signals as short as practical. Also, do not bundle (make a wiring harness) these wires. The problem you are trying to avoid is magnetic coupling (cross-talk).

6. Use a debounce circuit when setting up digital systems that require mechanical switches.

7. Connect all unused **TTL** *inputs* to the positive supply voltage through a 1k resistor. Or tie them to other similar inputs. It is extremely important that you realize that *every unused input pin* of a **CMOS IC** must be connected to either V_{DD} (the positive supply voltage) or V_{SS} (the negative supply voltage — generally ground).

8. Use a pull-up resistor for a larger high level output. Typically, a **TTL** gate's high level output pulse is 3.3. volts. You can increase this to + 5 volts by adding a 2.2k resistor between a + 5 volt source and the gate's output lead.

9. Use an output transistor if you need more current to drive a following unit. Place a 1k resistor between the base of the output transistor and the gate's output lead.

10. **CMOS** handling precautions

   a. Do not store **CMOS** chips in plastic trays, bags or foam.

   b. Always short the pins of a **MOS** chip. Wrap fine wire or, easier, wrap aluminum foil around the chip so all pins are shorted together.

   c. Do not touch the pins of **CMOS IC**'s.

   d. When working with **CMOS** chips, use **IC** sockets. Or, use a grounded soldering iron.

   e. Check pin configurations of **CMOS** chips before substituting them for a **TTL IC**. Some are pin-for-pin compatible, some are not! Example, you can not directly replace a **CMOS** 4011 (quad 2-input **NAND** gate) for a **TTL** 7400 (quad 2-input **NAND** gate) without rewiring the **IC** socket.

INDEX